Editor
Lorin Klistoff, M.A.

Managing Editor
Karen J. Goldfluss, M.S. Ed.

Editor-in-Chief
Sharon Coan, M.S. Ed.

Cover Artist
Barb Lorseyedi

Art Coordinator
Kevin Barnes

Art Director
CJae Froshay

Imaging
Alfred Lau
Ralph Olmedo, Jr.

Product Manager
Phil Garcia

Publishers
Rachelle Cracchiolo, M.S. Ed.
Mary Dupuy Smith, M.S. Ed.

Problem Solving Math Journals

for Intermediate Grades

$(2+y) - 9 = 3$

Author

Mary Rosenberg

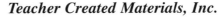

Teacher Created Materials, Inc.
6421 Industry Way
Westminster, CA 92683
www.teachercreated.com
ISBN-0-7439-3356-7
©2002 Teacher Created Materials, Inc.
Made in U.S.A.

Teacher Created Materials

Table of Contents

Introduction

As educators and parents it is imperative that each student understand how to use, to manipulate, to compare, to compute numbers easily and fluently, and to develop an understanding of number patterns, number relationships, and number functions. Much of the technology that is being used today and will be used in the future requires students to have a strong foundation of mathematics, to have the knowledge and expertise of knowing how and when to apply those skills in new situations, and to solve problems that require more than one- or two-step solutions.

Meeting the Math Standards

The activity pages in *Problem Solving Math Journals for Intermediate Grades* reinforce the math standards developed by the National Council of Teachers of Mathematics (NCTM) and provide the students with opportunities to practice and review important math skills. Through the use of this book, the student will be able to do the following:

- broaden his or her understanding of number sense
- add and subtract number to the billions
- multiply numbers with factors containing two or more digits
- divide numbers with remainders
- add, subtract, multiply, divide, compare like fractions, unlike fractions, mixed fractions, and improper fractions
- work with decimals, money, and percents
- solve math problems requiring two or more steps
- develop an understanding of probability
- work with statistics, geometry, time, and measurement

How to Use This Book

This book can be used in a wide variety of ways to best meet the needs of each student.

- Selected pages can be photocopied and assembled into a packet for each student. For example, photocopy all of the pages that work with a specific skill or concept, such as time, money, or fractions.
- For whole-class instruction, complete a page each day as part of the morning routine or as a warm-up for that day's math lesson.

- Selected pages can be placed at a center, used for seatwork, sent home for homework, or to review/preview new concepts and skills.

At the bottom of each activity page, there is a section for students to use. Students can create and solve their own math problem and then explain their solution. This section is optional and can be used as an extension. The teacher can place these student-created problems into new worksheets or a problem-solving book.

Developing a Math Vocabulary

It is important that each student develop a comprehensive math vocabulary using terms that are unique to math and are used or read about on a regular basis. For example, the student might read a newspaper article about a player's batting average going up or down, or the student might listen to a radio report on the rise and fall of the stock market. To be able to effectively judge and evaluate the information for accuracy and relevance, the student needs to independently compare the information to the facts, have an understanding of the reasonableness of a conclusion or fact, and know how that same data can be manipulated to paint a rosier or gloomier picture.

This book provides a "Math Vocabulary" (pages 239–256). Each page of the math dictionary contains a math-related vocabulary term, its definition, and provides a space for the student to write or draw an example problem. Each vocabulary word is ordered in the same sequence that the word is used in the math journal pages.

To use the math dictionary, photocopy one set of words for each student. Provide each student with a small blank journal. When a new word is used in a word problem, have students cut out the appropriate word card, write and/or draw an example problem, and glue the page in the math journal. By having the student write or draw an example problem, the student is able to show his or her understanding of the word, concept, or skill. As more words are added to the dictionary, the student is building a great reference tool that can be used at home, in the classroom, or in future math classes.

Name: _____ Date: _____

Warm Up

Write the unit of time each event lasts—seconds, minutes, or hours.

A. a birthday party	B. a blink of an eye	C. eating a sandwich	D. a sneeze	E. cooking a turkey
_____	_____	_____	_____	_____

Word Problem

F. Read each clue. If the answer is "yes" make an "O" in the box. If the answer is "no" make an "X" in the box. (*Note:* The houses are in order from shortest to longest building time.)

- Ivan's house didn't take the most nor the least amount of time to build.
- Fay is allergic to wood so she would never use it to build a house.
- Gary likes to use wood in every house he makes.
- Helen likes how long brick houses last.

Who built each home?

sand: _____ blocks: _____ bricks: _____ tree: _____

	Fay	Gary	Helen	Ivan
wood blocks				
sand castle				
tree house				
bricks				

Create your own math problem and explain your solution.

Math Problem:

Explanation/Solution:

_____

Name: _____ **Date:** _____

Warm Up

Circle the operation.

A. Ali's team scored 5 points in the first half and 2 points in the second half. How many points in all?

add   subtract

B. Cori made 12 aces in the first set and 6 aces in the second set. How many aces in all?

add   subtract

C. Danny made 9 slam dunks yesterday and 4 fewer slam dunks today. How many slam dunks did Danny make today?

add   subtract

Word Problem

D. The pep squad scored 83 points for giving the loudest cheer. They scored 27 fewer points for standing in straight rows and columns. How many points did they score for standing in straight rows and columns?

Create your own math problem and explain your solution.

Math Problem:

Explanation/Solution:

Name: _____ Date: _____

3

Warm Up

Add.

A.

$2 + 5 =$ _____

B.

$8 + 9 =$ _____

C.

$3 + 6 =$ _____

D.

$8 + 7 =$ _____

E.

$6 + 3 =$ _____

F.

$4 + 6 =$ _____

G.

$5 + 1 =$ _____

H.

$2 + 1 =$ _____

I.

$5 + 4 =$ _____

J.

$5 + 10 =$ _____

Word Problem

K. Melanie planted 8 petunias in one row and 3 snapdragons in another row.

- How many plants did Melanie plant? _____
- How many rows of plants did Melanie plant? _____

Create your own math problem and explain your solution.

Math Problem:

Explanation/Solution: _____

Warm Up

Name: _____ **Date:** _____

Write the missing number.

A.

$5 +$ _____ $= 15$

B.

$9 +$ _____ $= 17$

C.

$4 +$ _____ $= 7$

D.

_____ $+ 5 = 9$

E.

$1 + 16 =$ _____

F.

$20 -$ _____ $= 17$

G.

$16 - 10 =$ _____

H.

_____ $- 7 = 11$

I.

$13 -$ _____ $= 8$

J.

$14 -$ _____ $= 9$

Word Problem

K. Meredith had 18¢. She found 4¢ in her pocket. She found the rest of the money under her bed. How much money was under her bed?

Create your own math problem and explain your solution.

Math Problem:

Explanation/Solution: _____

Name: _____

Date: _____

Warm Up

Find the perimeter for each shape.

A.

_____ units

B.

_____ units

C.

_____ units

D.

_____ units

Word Problem

E. Draw a square with a perimeter of 20 units.

F. Draw a rectangle with a perimeter of 20 units.

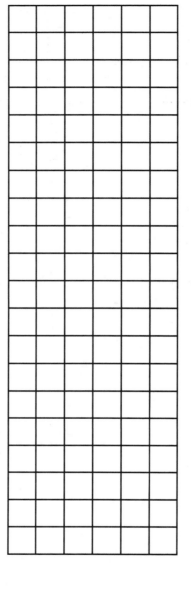

Create your own math problem and explain your solution.

Math Problem: _____

Explanation/Solution: _____

Answers: A. 16 units B. 14 units C. 14 units D. 10 units E. Each side of the square will be 5 units. F. Answers will vary: 1 unit by 20 units or 2 units by 10 units or 4 units by 5 units.

Warm Up

Subtract.

A.

$11 - 10 =$ _____

B.

$18 - 5 =$ _____

C.

$19 - 13 =$ _____

D.

$15 - 4 =$ _____

E.

$14 - 9 =$ _____

F.

$12 - 6 =$ _____

G.

$16 - 5 =$ _____

H.

$17 - 6 =$ _____

I.

$17 - 9 =$ _____

J.

$13 - 8 =$ _____

Word Problem

1. How many more pumpkins did Stacy grow than Tracy? _____
2. Who grew more carrots? _____
3. How many fewer watermelons did Tracy grow than Stacy? _____
4. What was the total number of plants Stacy grew? _____
5. What was the total number of plants Tracy grew? _____
6. What was the difference between the total numbers? _____

K.

	Stacy	Tracy
pumpkins	19	10
strawberries	5	8
watermelons	18	11
carrots	12	16

Create your own math problem and explain your solution.

Math Problem:

Explanation/Solution:

Name: _____ Date: _____

Warm Up

Count the money. Write the amount using a dollar sign and decimal point.

A.  _____

B.  _____

C.  _____

D.  _____

E.  _____

Word Problem

F. Using the fewest number of coins, identify which coins each customer used to pay for one of the dog services.

Doggy Business	
Daily Walking	$0.35
Grooming	$0.67
Bathing	$1.94
Play Fetch	$0.81
Flea Check	$0.20

1. Mrs. Brown had Fifi checked for fleas. _____

2. Mr. Green had Bruno groomed. _____

3. Mrs. White had Fido take a bath. _____

4. Mr. Black had Dudley play fetch. _____

5. Mrs. Gray had Buster taken for a walk. _____

Create your own math problem and explain your solution.

Math Problem:

_____

Explanation/Solution:

_____

Answers: A. $0.12 B. $0.25 C. $1.50 D. $0.75 E. $0.90 F1. 2 dimes F2. 1 half dollar, 1 dime, 1 nickel, 2 pennies F3. 3 half dollars, 1 quarter, 1 dime, 1 nickel, 4 pennies F4. 1 half dollar, 1 quarter, 1 nickel, 1 penny F5. 1 quarter, 1 dime

10

Warm Up

Name: _____ **Date:** _____

Compare the money using the signs > or <.

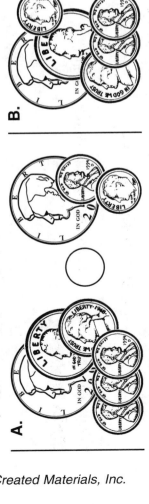

A. ◯ B. ◯ C.

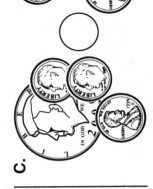

😎

Word Problem

D. Nathan had $1.00. Each transaction at the change machine costs $0.15. If Nathan used the change machine 3 times, how much money would he have left?

Create your own math problem and explain your solution.

Math Problem:

Explanation/Solution: _____

Name:

Date:

Warm Up

Write the number.

A. number of nickels in $1.00	**B.** number of half dollars in $1.00	**C.** number of quarters in $1.00	**D.** number of pennies in $1.00	**E.** number of dimes in $1.00
_____	_____	_____	_____	_____

Word Problem

F. Which is worth more: 77 nickels or 47 dimes?

Create your own math problem and explain your solution.

Math Problem:

Explanation/Solution:

_____

Answers: A. 20 B. 2 C. 4 D. 100 E. 10 F. 47 dimes are worth more. 77 nickels = $3.85, 47 dimes = $4.70

Name: _____ **Date:** _____

Warm Up

Find the equivalent amount.

A. one quarter =
_____ pennies

B. one half dollar =
_____ dimes

C. two dimes =
_____ nickels

D. one nickel =
_____ pennies

E. two half dollars =
_____ quarters

F. three quarters =
_____ nickels

G. twenty pennies =
_____ dimes

H. 10 nickels =
_____ half dollar

I. fifty pennies =
_____ quarters

J. four nickels =
_____ dimes

Word Problem

K. Miguel started with the money shown below, and he spent $1.54 at the Model Airplane Shop. How much money does Miguel have left?

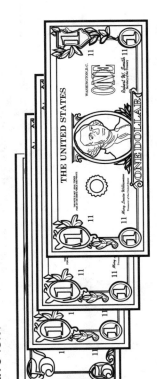

Create your own math problem and explain your solution.

Math Problem:

Explanation/Solution:

Warm Up

Write each amount using a dollar sign and decimal point.

A. five dollars and eight-five cents

B. nine dollars and fifty cents

C. twenty-two cents

D. eighteen dollars and eleven cents

E. forty-one dollars and sixty-one cents

Word Problem

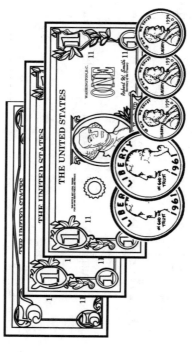

F. 1. Count the money.

2. What could you buy with the money? Make a list.

Create your own math problem and explain your solution.

Math Problem:

Explanation/Solution:

Name: _____

Date: _____

Warm Up

A **ratio** tells the mathematical relationship between two items. Write the ratio two ways.

Example:
1 car to 4 tires
1:4 or 1/4

A. dog's tail to paws _____

B. eyes to fingers _____

C. bike to wheels _____

D. months to year _____

Word Problem

E. Write the ratio two ways.

1. Jamilah earns $50.00 for each computer she repairs. _____

2. Jamilah has one day off for every nine days she works. _____

3. Each computer has three cables. _____

4. There are 104 keys on one keyboard. _____

5. The special offer is buy two disks and receive one for free. _____

Create your own math problem and explain your solution.

Math Problem:

_____

Explanation/Solution:

_____

Answers: A. 1:4 or 1/4 B. 2:10 or 2/10 C. 1:2 or 1/2 D. 12:1 or 12/1 E1. $50.00 to 1, 50:1, or 50/1 E2. 1:9 or 1/9 E3. 1:3 or 1/3 E4. 104:1 or 104/1 E5. 2:1 or 2/1

13

Name: _____

Date: _____

Warm Up

Solve.

A.
$(10 + 2) + 9 =$ _____

B.
$4 + (3 + 1) =$ _____

C.
$(7 + 6) + 3 =$ _____

D.
$4 + (4 + 3) =$ _____

E.
$(4 + 7) + 3 =$ _____

F.
$2 + (7 + 3) =$ _____

G.
$(10 + 3) + 2 =$ _____

H.
$3 + (8 + 6) =$ _____

I.
$(4 + 2) + 2 =$ _____

J.
$6 + (3 + 8) =$ _____

Word Problem

K. Write the problem. Solve.

One day Ira picked 3 tomato plants and 7 potato plants. The next day Ira picked 4 heads of lettuce. How many vegetables did Ira pick in all?

Create your own math problem and explain your solution.

Math Problem:

Explanation/Solution: _____

Answers: A. 21 B. 8 C. 16 D. 11 E. 14 F. 12 G. 15 H. 17 I. 8 J. 17 K. (3 + 7) + 4 = 14 vegetables

14

Warm Up

Write "exact" or "estimate."

A. The time is 3:24.

_____

B. Mercury is about 36 million miles away from the sun.

_____

C. The baby weighs about ten pounds.

_____

D. The address is 123 Main Street.

_____

E. My birthday is about 6 months away.

_____

Word Problem

F. Add the numbers two ways—with an exact answer and with an estimated answer (round the numbers first to the nearest dollar). Compare the two answers.

Exact	Estimate
$99.10	_____
$18.14	_____
+ $33.86	_____

Create your own math problem and explain your solution.

Math Problem:

Explanation/Solution:

_____

15

Name: _____ **Date:** _____

Warm Up

Complete each subtraction table.

A.

−	48	16	55	32	29	31	10	8
5								

B.

−	16	67	31	78	17	45	24	59
10								

Word Problem

C. Ana sold 22 pairs of pants. She sold 6 fewer pairs of tops and 1 more sweater. How many tops and sweaters did Ana sell in all?

Create your own math problem and explain your solution.

Math Problem:

Explanation/Solution:

_____

Warm Up

Rewrite each problem rounding each number to the nearest ten. Solve.

A.

24 – 20 = _____

_____

B.

58 + 8 = _____

_____

C.

86 – 72 = _____

_____

D.

62 + 43 = _____

_____

Word Problem

E. Round each item and then solve the problem.

Grandpa's candy jar can hold 100 pieces of candy. Select the most number of different kinds of candies that will fit in the candy jar.

Candies	
kinds of candies	**rounded**
16 jellybeans	_____
27 bite-size candy bars	_____
19 sour balls	_____
71 chocolate pops	_____
36 suckers	_____

Create your own math problem and explain your solution.

Math Problem: _____

Explanation/Solution:

Answers: A. 20 – 20 = 0 B. 60 + 10 = 70 C. 90 – 70 = 20 D. 60 + 40 = 100 E. candies: 20, 30, 20, 70, 40; jellybeans, bite-size candy bars, sour balls, and suckers

Name: _____   **Date:** _____

Warm Up **17**

Write the time in standard form.

A.	B.	C.	D.	E.
one twenty-nine	three twenty-six	half past three	quarter past two	quarter until 7
_____	_____	_____	_____	_____

Word Problem

F. Fill in the chart and answer the questions.

Bus	Departs	Arrives	Elapsed Time
1	3:45 P.M.	3:51 P.M.	_____
2	2:36 P.M.	2:48 P.M.	_____
3	1:12 P.M.	1:43 P.M.	_____
4	10:09 A.M.	10:31 A.M.	_____
5	6:10 P.M.	6:27 P.M.	_____

1. Which bus has the longest travel time? _____
2. Which bus has the shortest travel time? _____
3. Which bus runs in the morning? _____
4. Which buses run in the afternoon? _____
5. Which bus departs first? _____
6. Which bus departs last? _____
7. Which bus would arrive around 4:00 P.M.? _____

Create your own math problem and explain your solution.

Math Problem:

Explanation/Solution:

Name: _____

Date: _____

Warm Up

Rewrite each time in standard form using A.M. and P.M.

A.	**B.**	**C.**	**D.**	**E.**
eighteen-hundred hours	twenty-three-hundred hours	thirteen-hundred hours	sixteen-hundred hours	zero one hundred hours
_____	_____	_____	_____	_____

Word Problem

F. The International Cuisine Restaurant is open from ten-hundred hours to twenty-two-hundred hours. Will Sierra be able to pick up her take-out order at 1:00 P.M.? When does this restaurant open and close in standard form (civilian time)?

Create your own math problem and explain your solution.

Math Problem:

Explanation/Solution: _____

Name: _____ **Date:** _____

Warm Up

If it is noon in New York, what is the time in other parts of the world? Write the time using A.M. and P.M.

A. It is the same time in Akron, Ohio.

___ : ___

B. It is three hours earlier in Portland, Oregon.

___ : ___

C. It is six hours later in Geneva, Switzerland.

___ : ___

D. It is nine hours earlier in Melbourne, Australia.

___ : ___

E. It is five hours later in London, England.

___ : ___

0   +1   +2   +3

Word Problem

F. Kelsey leaves Los Angeles at 9:00 A.M. and flies for three hours before arriving in New York. What is the time in New York when she lands?

Create your own math problem and explain your solution.

Math Problem:

Explanation/Solution:

_____

Answers: A. 12:00 P.M. B. 9:00 A.M. C. 6:00 P.M. D. 3:00 A.M. E. 5:00 P.M. F. The plane will arrive at 3:00 P.M.

Name: _____ **Date:** _____

Warm Up

Circle the holiday that comes next.

A.	**B.**	**C.**	**D.**	**E.**
Independence Day	Father's Day	New Year's Day	Kwanzaa	Labor Day
Flag Day	Mother's Day	New Year's Eve	Hanukkah	Memorial Day

Word Problem

F. Use a calendar (a current one or an out-of-date one) to make a list of all of the holidays in one year and the day on which they land. (A holiday is a "no school day.")

Holiday	Day	Holiday	Day	Holiday	Day

Create your own math problem and explain your solution.

Math Problem:

_____

Explanation/Solution:

_____

Answers: A.–E.: Answers will vary. F. Answers may vary. New Year's Day (Jan. 1), Christmas (Dec. 25), Thanksgiving, Veteran's Day, Memorial Day, Labor Day, Good Friday/Easter

Name: _____ **Date:** _____

Warm Up

In a bag there are these items. Circle the one that would most likely be picked.

A. 10 pennies and 1 dime	**B.** 20 red crayons and 10 blue crayons	**C.** 8 peanuts and 80 walnuts	**D.** 13 red counters, 30 blue counters, and 3 orange counters	**E.** 9 brown beans, 29 white beans, and 15 red beans
penny dime	red crayon bule crayon	peanut walnut	red blue orange	brown white red

Word Problem

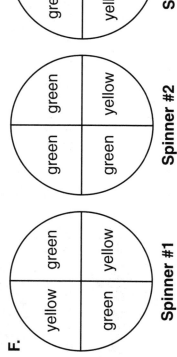

yellow	green		green	green		green	yellow
green	yellow		green	yellow		yellow	yellow

Spinner #1   **Spinner #2**   **Spinner #3**

1. Which spinner would be more likely to land on green? _____ Why? _____

2. Which spinner would be more likely to land on yellow? _____ Why? _____

3. Which spinner would have an equal chance of landing on green or yellow? _____

   Why? _____

Create your own math problem and explain your solution.

Math Problem:

Explanation/Solution:

Answers: A. penny B. red crayon C. walnut D. blue counter E. white bean F1. #2 because it has more green sections than yellow sections. F2. #3 because it has more yellow sections than green sections. F3. #1 because it has the same number of green and yellow sections.

22

Warm Up

Make two addition and two subtraction facts using each set of numbers.

A. 6, 7, 13

___ + ___ = ___
___ + ___ = ___
___ − ___ = ___
___ − ___ = ___

B. 2, 7, 9

___ + ___ = ___
___ + ___ = ___
___ − ___ = ___
___ − ___ = ___

C. 6, 9, 15

___ + ___ = ___
___ + ___ = ___
___ − ___ = ___
___ − ___ = ___

D. 6, 12, 18

___ + ___ = ___
___ + ___ = ___
___ − ___ = ___
___ − ___ = ___

Word Problem

E. Billy had 14 keys. Three of the keys were house keys. The rest of the keys were car keys. How many of the keys were car keys?

Create your own math problem and explain your solution.

Math Problem: _____

Explanation/Solution: _____

Answers: A. $6 + 7 = 13$, $7 + 6 = 13$, $13 − 6 = 7$, $13 − 7 = 6$ *B.* $2 + 7 = 9$, $7 + 2 = 9$, $9 − 2 = 7$, $9 − 7 = 2$ *C.* $6 + 9 = 15$, $9 + 6 = 15$, $15 − 9 = 6$, $15 − 6 = 9$ *D.* $6 + 12 = 18$, $12 + 6 = 18$, $18 − 12 = 6$, $18 − 6 = 12$ *E.* $14 − 3 = 11$ *car keys*

Name: _____

Date: _____

Warm Up

Add.

A.

$$\frac{1}{4} + \frac{1}{4} = \underline{\quad}$$

B.

$$\frac{5}{12} + \frac{3}{12} = \underline{\quad}$$

C.

$$\frac{9}{20} + \frac{5}{20} = \underline{\quad}$$

D.

$$\frac{11}{16} + \frac{4}{16} = \underline{\quad}$$

E.

$$\frac{3}{15} + \frac{10}{15} = \underline{\quad}$$

Word Problem

F. Freddy mixed together 1/3 cup of white sugar with 1/3 cup of brown sugar. How much sugar did Freddy mix together?

Create your own math problem and explain your solution.

Math Problem:

Explanation/Solution: _____

Name: _____ **Date:** _____

Warm Up

Add.

A.
$$\frac{2}{6} + \frac{3}{6} = \underline{\quad}$$

B.
$$\frac{4}{7} + \frac{1}{7} = \underline{\quad}$$

C.
$$\frac{1}{8} + \frac{3}{8} = \underline{\quad}$$

D.
$$\frac{1}{6} + \frac{3}{6} = \underline{\quad}$$

E.
$$\frac{2}{5} + \frac{2}{5} = \underline{\quad}$$

Word Problem

F. Jan ate 1/5 of the pizza and 2/5 of the pie. How much food did Jan eat?

Create your own math problem and explain your solution.

Math Problem:

Explanation/Solution: _____

Answers: A. 5/6 B. 5/7 C. 4/8 D. 4/6 E. 4/5 F. 1/5 + 2/5 = 3/5 of the food

25

Warm Up

Subtract.

A.
$$\frac{6}{10} - \frac{3}{10} = \underline{\qquad}$$

B.
$$\frac{7}{8} - \frac{5}{8} = \underline{\qquad}$$

C.
$$\frac{3}{4} - \frac{2}{4} = \underline{\qquad}$$

D.
$$\frac{6}{9} - \frac{5}{9} = \underline{\qquad}$$

E.
$$\frac{7}{8} - \frac{2}{8} = \underline{\qquad}$$

Word Problem

F. Fola's gas tank was 4/7 filled. She used 2/7 of the gas running errands. How much gas does Fola have left in her car?

Create your own math problem and explain your solution.

Math Problem:

Explanation/Solution: _____

Answers: A. 3/10 B. 2/8 C. 1/4 D. 1/9 E. 5/8 F. 4/7 – 2/7 = 2/7 tank of gas

26

Warm Up

Add the missing + or − signs.

A.   9 ___ 9 ___ 8 = 10

B.   4 ___ 4 ___ 9 = 17

C.   8 ___ 3 ___ 10 = 15

D.   5 ___ 5 ___ 0 = 0

E.   7 ___ 7 ___ 1 = 13

F.   5 ___ 3 ___ 3 = 5

G.   5 ___ 7 ___ 10 = 2

H.   5 ___ 1 ___ 6 = 0

I.   1 ___ 4 ___ 9 = 14

J.   9 ___ 8 ___ 6 = 7

Word Problem

K. Write the math problem. Solve.

Julie sold 7 pairs of sneakers and 7 pairs of sandals. Four pairs of sneakers were returned. What was the total number of shoes sold by Julie?

Create your own math problem and explain your solution.

Math Problem:

Explanation/Solution:

_____

Answers: A. 9 + 9 − 8 = 10 *B.* 4 + 4 + 9 = 17 *C.* 8 − 3 + 10 = 15 *D.* 5 − 5 + (or −) 0 = 0 *E.* 7 + 7 − 1 = 13 *F.* 5 + 3 − 3 = 5 *G.* 5 + 7 − 10 = 2 *H.* 5 + 1 − 6 = 0 *I.* 1 + 4 + 9 = 14 *J.* 9 − 8 + 6 = 7 *K.* 7 + 7 − 4 = 10 *shoes*

Name:

Date:

Warm Up

Circle the event that is more likely to occur.

A.

_____

It will rain today.

It will not rain today.

B.

The sun will shine.

An alien will come to class.

C.

Mom will see a robin.

Mom will see a dodo bird.

D.

_____

Finding a penny in a pocket.

Finding a $100 in a pocket.

Word Problem

E. Mrs. Silverberg said there would be a pop quiz sometime this week. It is Thursday afternoon, and school has ended. The students have not had a pop quiz. What are the chances that the students will be given a pop quiz on Friday? What makes you think that?

Create your own math problem and explain your solution.

Math Problem: _____

Explanation/Solution: _____

Answers: A. Answers will vary. B. The sun will shine. C. Mom will see a robin. D. Finding a penny in a pocket. E. The test will be given on Friday because that is the last school day for the week.

28

Name:

Date:

Warm Up

Circle the event most likely to occur today.

A.

going to the movies

being in the movies

watching a movie in class

B.

seeing a car

driving a race car

building a car

C.

painting a house

painting a picture

painting a room

D.

catching a bus

catching a fish

catching a kangaroo

Word Problem

E.

1. Shuffle a deck of cards and place in a stack facedown on the table. Which cards will most likely appear—face cards or numbered cards? Why?

2. Turn over the first ten cards. Use tally marks to record what kind of card was turned over. Compare the results to your guess.

Face Cards	
Numbered Cards	

Create your own math problem and explain your solution.

Math Problem:

Explanation/Solution:

Answers: A.–D. Answers will vary. E1. More likely to turn over a numbered card because there are more numbered cards than face cards. E2. Chart will vary.

29

Name: _____

Date: _____

Warm Up

Write the place value for the underlined digit.

A.

69<u>1</u>

_____

B.

2<u>5</u>1

_____

C.

<u>4</u>,913

_____

D.

2,<u>8</u>47

_____

E.

3,0<u>4</u>6

_____

Word Problem

F. Rewrite the problem in standard form. Solve.

Warren read four hundred seventy-three pages. Nancy read five hundred twenty-one pages. How many pages did they read in all?

Create your own math problem and explain your solution.

Math Problem: _____

Explanation/Solution: _____

3

Warm Up

Cross out the unnecessary piece of information.

A. Deb had $0.64. Three of the coins were shiny. Deb found another $0.18. How much money does Deb have?

B. Jeb's sister had $0.47. Jeb had $0.92. He spent $0.10 buying candy. How much money does Jeb have left?

C. Caleb had $0.87. He earned $0.39 doing housework. Caleb loves to clean. How much money does he now have?

D. Wilma went to the movies. She spent $0.28 on a soda and $0.49 on the ticket. How much did Wilma spend?

Word Problem

E. Which piggy bank is worth the most amount of money? How do you know that?

Piggy Bank #1	Piggy Bank #2	Piggy Bank #3
4 pennies	1 penny	3 pennies
3 nickels	2 nickels	2 nickels
2 dimes	3 dimes	4 dimes
1 quarter	4 quarters	1 quarter

Create your own math problem and explain your solution.

Math Problem:

Explanation/Solution:

_____

Answers: A. Three of the coins were shiny. B. Jeb's sister had $0.47. C. Caleb loves to clean. D. Wilma went to the movies. E. Piggy Bank #2 is worth the most amount of money; Piggy Bank #1: $0.64, Piggy Bank #2: $1.41, Piggy Bank #3: $0.78

Answers: A. 210 B. 240 C. 230 D. 210 E1. 140 E2. 170 E3. 0 E4. 30 E5. 70 E6. 60

Warm Up

Rewrite each problem rounding each number to the nearest ten.   Add.

A.
```
   91   _____
   19   _____
   78   _____
 + 24   + _____
_____  _____
```

B.
```
   46   _____
   64   _____
   73   _____
 + 62   + _____
_____  _____
```

C.
```
   35   _____
   47   _____
   51   _____
 + 89   + _____
_____  _____
```

D.
```
   46   _____
   71   _____
   12   _____
 + 82   + _____
_____  _____
```

Word Problem

E. Write the math problem rounding each number to the nearest ten.

| milk balls 53 | taffy 77 | jellybeans 92 |
| fireballs 15 | licorice 24 | bubble gum 93 |

1. jellybeans + milk balls

_____ + _____ = _____

2. taffy + bubble gum

_____ + _____ = _____

3. licorice – fireballs

_____ – _____ = _____

4. milk balls – licorice

_____ – _____ = _____

5. jellybeans – fireballs

_____ – _____ = _____

6. taffy – licorice

_____ – _____ = _____

Create your own math problem and explain your solution.

Math Problem: _____

Explanation/Solution: _____

Name: _____

Date: _____

Warm Up

Write each number in standard form.

A.

three hundred seventy-two

_____

B.

five hundred fifty-one

_____

C.

nine hundred sixty-four

_____

D.

eight hundred seven

_____

Word Problem

E. If Michaela left Vista Valley and drove to Cold River and from Cold River to Scottstown, how many miles would Michaela have to drive?

Vista Valley to:	
Scottstown	105 miles
Mountain Valley	258 miles
Oakville	322 miles
Cold River	437 miles

Create your own math problem and explain your solution.

Math Problem:

_____

Explanation/Solution:

_____

Answers: A. 372 B. 551 C. 964 D. 807 E. 437 miles + 332 miles (Cold River – Scottstown) = 769 miles

Name: _____ Date: _____

33

Warm Up

Add.

A.
```
  234
  355
+ 310
_____
```

B.
```
  411
  223
+ 352
_____
```

C.
```
  341
  321
+ 135
_____
```

D.
```
  334
  234
+ 211
_____
```

E.
```
  154
  405
+ 340
_____
```

Word Problem

F. Jade inventoried the bookstore. She counted 521 children's books, 234 bibliographies, and 133 science-fiction novels. How many books did Jade count?

Create your own math problem and explain your solution.

Math Problem:

Explanation/Solution:

Name: _____

Date: _____

Warm Up

Add.

A.
```
  313
  546
+ 134
_____
```

B.
```
  508
  109
+ 259
_____
```

C.
```
  721
  139
+ 106
_____
```

D.
```
  208
  613
+ 102
_____
```

E.
```
  541
  236
+ 107
_____
```

Word Problem

F. On Monday, Mrs. Oswald checked out 215 books. On Tuesday and Wednesday, Mrs. Oswald checked out 317 books and 429 books. How many books did Mrs. Oswald check out in three days?

Create your own math problem and explain your solution.

Math Problem:

Explanation/Solution: _____

Warm Up

Name: _____

Date: _____

Add.

A.
```
  410
  359
+ 171
_____
```

B.
```
  601
  199
+ 154
_____
```

C.
```
  564
  272
+ 119
_____
```

D.
```
  158
  307
+ 453
_____
```

E.
```
  121
  103
+ 496
_____
```

Word Problem

F. How many copies of the top three books were sold?

Best Sellers

1. Bed Head and the Three Bears	527 copies
2. Little Blue Ball Cap	238 copies
3. Rumpleheadroom	196 copies
4. Hans and Greta	175 copies
5. Princess and the Peabodys	106 copies

Create your own math problem and explain your solution.

Math Problem: _____

Explanation/Solution: _____

36

Warm Up

Write an event.

A. More likely to occur today:

_____
_____
_____

B. Least likely to occur today:

_____
_____
_____

C. Might occur today:

_____
_____
_____

D. Would never occur today:

_____
_____
_____

E. Would definitely occur today:

_____
_____
_____

Word Problem

F. Take a coin and flip it 20 times in a row. Use tally marks to record how the coin lands. Using a different coin, flip it 20 times and use tally marks to record how the coin lands. Compare the results between the two different coins.

Coin #1

Heads	
Tails	

Coin #2

Heads	
Tails	

Create your own math problem and explain your solution.

Math Problem:

Explanation/Solution: _____

Answers: A–F. Answers will vary.

Name: _____

Date: _____

Warm Up

Use the map to answer the questions.

A. Circle the longer trip.

Seacrest to Ocean View

Seacrest to Ocean Spray

B. Circle the longer trip.

Ocean Spray to Mt. View

Ocean Spray to Ocean View

C. Circle the shorter trip.

Waverly to High Point

Waverly to Ocean Spray

D. Circle the shorter trip.

High Point to Seacrest

High Point to Ocean Spray

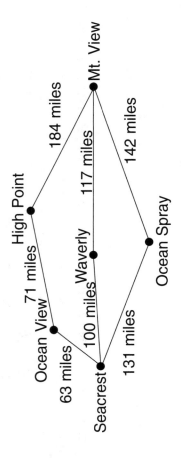

Word Problem

E. Find the shortest distance from Ocean View to Mt. View.

Create your own math problem and explain your solution.

Math Problem:

Explanation/Solution:

Warm Up

Subtract.

A.
```
   468
 - 133
 _____
```

B.
```
   349
 - 345
 _____
```

C.
```
   654
 - 432
 _____
```

D.
```
   831
 - 520
 _____
```

E.
```
   950
 - 730
 _____
```

Word Problem

F. Jarrod read 915 pages for the reading marathon. Jasper read 813 fewer pages than Jarrod. How many pages did Jasper read?

_____

Create your own math problem and explain your solution.

Math Problem:

Explanation/Solution:

_____

Answers: A. 335 B. 4 C. 222 D. 311 E. 220 F. 102 pages

39

Name: _____

Date: _____

Warm Up

Add or subtract.

A.
```
  493
+ 441
_____
```

B.
```
  310
+ 518
_____
```

C.
```
  723
- 614
_____
```

D.
```
  846
- 462
_____
```

E.
```
  773
- 108
_____
```

Word Problem

F. Hans read 159 pages on Friday evening. He read 45 fewer pages on Saturday and 164 more pages on Sunday. How many pages did Hans read over the weekend?

Create your own math problem and explain your solution.

Math Problem:

Explanation/Solution:

Warm Up

40

Subtract.

A.
```
  810
- 120
_____
```

B.
```
  415
- 160
_____
```

C.
```
  877
- 369
_____
```

D.
```
  619
- 537
_____
```

E.
```
  721
- 404
_____
```

Word Problem

F. Hubert went to the library and checked out a book. The book had 197 pages. Hubert read 88 pages. How many more pages does Hubert have left to read?

Create your own math problem and explain your solution.

Math Problem:

Explanation/Solution:

Answers: A. 690 B. 255 C. 508 D. 82 E. 317 F. 109 pages

Name: _____

Date: _____

Warm Up

Do the operation in parentheses first.

A.

$(74 + 79) - 86 =$ _____

B.

$10 + (73 - 15) =$ _____

C.

$(83 - 67) + 68 =$ _____

D.

$(51 - 10) + 86 =$ _____

E.

$99 + (95 - 21) =$ _____

Word Problem

F. Write the equation. Solve.

Jacinda had $0.76 and her sister had $10.68. They spent $5.40. How much money do the sisters have left?

_____

Create your own math problem and explain your solution.

Math Problem:

Explanation/Solution: _____

Answers: *A. 153 − 86 = 67 B. 10 + 58 = 68 C. 16 + 68 = 84 D. 41 + 86 = 127 E. 99 + 74 = 173 F. ($0.76 + $10.68) − $5.40 = $11.44 − $5.40 = $6.04*

Name: _____ **Date:** _____

Warm Up

Rewrite each problem rounding each number to the nearest hundred. Add.

A.
```
  710  _____
  491  _____
  672  _____
+ 331  + _____
       _____
```

B.
```
  923  _____
  148  _____
  781  _____
+ 324  + _____
       _____
```

C.
```
  591  _____
   31  _____
   58  _____
+ 265  + _____
       _____
```

D.
```
  588  _____
   99  _____
  339  _____
+ 108  + _____
       _____
```

Word Problem

Sport Participants

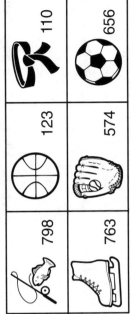

798	123	110
763	574	656

E. Write the problem rounding each number to the nearest hundred.

1. fishing + baseball

   _____ + _____ = _____

2. karate + soccer

   _____ + _____ = _____

3. skating + basketball

   _____ + _____ = _____

4. baseball – karate

   _____ – _____ = _____

5. skating – soccer

   _____ – _____ = _____

6. fishing – basketball

   _____ – _____ = _____

Create your own math problem and explain your solution.

Math Problem: _____

Explanation/Solution: _____

Name: _____   **Date:** _____

Warm Up

Rewrite each math problem rounding each number to the nearest hundred. Solve.

A.

981 − 325 = _____

_____

B.

810 − 759 = _____

_____

C.

681 + 326 = _____

_____

D.

410 + 164 = _____

_____

Word Problem

E. Round each number to the nearest hundred. Solve the problem.

Janice had 875 beads. She used 72 beads to make a bracelet and 362 beads to make a necklace. About how many beads does Janice have left?

Create your own math problem and explain your solution.

Math Problem:

Explanation/Solution: _____

Answers: A. 1,000 − 300 = 700 B. 800 − 800 = 0 C. 700 + 300 = 1,000 D. 400 + 200 = 600 E. 900 − 100 − 400 = 400 beads

44

Warm Up

Do the operation in parentheses first.

A.

$355 + (914 - 824) =$

_____

B.

$(223 + 772) - 536 =$

_____

C.

$(661 - 41) + 834 =$

_____

D.

$324 + (691 - 481) =$

_____

E.

$137 + (810 - 561) =$

_____

Word Problem

F. Write the equation. Solve.

Seabrook harvested 321 pounds of walnuts and 173 pounds of peanuts. He sold 218 pounds of nuts at the Town Harvest Festival. How many pounds of nuts does Seabrook have left?

Create your own math problem and explain your solution.

Math Problem:

Explanation/Solution: _____

Answers: A. $355 + 90 = 445$ B. $995 - 536 = 459$ C. $620 + 834 = 1,454$ D. $324 + 210 = 534$ E. $137 + 249 = 386$ F. $(321 + 173) - 218 = 494 - 218 = 276$ pounds of nuts

Warm Up

Name: _____

Date: _____

Circle the largest number. Draw a line under the smallest number.

A.		B.		C.		D.	
1,841		9,101		54,462		32,197	
1,716		2,827		93,601		55,448	
6,687		7,191		84,471		91,629	

Word Problem

E.

Rivers	
St. Lawrence	760 miles
Mississippi	2,348 miles
Columbia	1,210 miles
Yukon	1,979 miles
Snake	1,083 miles

1. Write the rivers in order, from shortest to longest.

   _____ , _____ , _____ , _____ , _____

2. What is the difference in miles between the longest river and the shortest river?

   _____

Create your own math problem and explain your solution.

Math Problem:

Explanation/Solution:

Name: _____

Date: _____

Warm Up

Answer each question.

A. Justin has earned the following test scores in math: F, D, C, D. Do you think Justin will get an "A" on the next test? Why or why not?

_____

B. Sarah has had 8 hits out of 10 at bats. Do you think Sarah will get a hit at her next at bat? Why or why not?

_____

Word Problem

C. Luigi Legumes is trying to get a hit movie—if only he can break into the top 100! Luigi has a new movie coming out. It is called *The Ship of the High Seas*. Do you think this movie will be a hit? _____ Why or why not? _____

Luigi's Movie Titles	Ranking
The Boat Won't Float	#996/1,000
The Ship's Name Is Rita	#999/1,000
My Dad Is a Boat	#1,000/1,000
The Boat Won't Go	#998/1,000
Who Took My Boat?	#995/1,000

(*Note:* The movies are ranked from 1 to 1,000. A ranking of #1 means that the movie was seen by the largest number of people for that week.)

Create your own math problem and explain your solution.

Math Problem:

Explanation/Solution: _____

Answers: A. No, Justin will not receive an "A" on the next test. He has not received an "A" or a "B" on any test. B. Yes, Sarah will get a hit because she has had hits for most of her other at bats. C. Luigi will not have a hit. None of his other movies ranked higher than 995.

Warm Up

47

Complete each table.

A.

+ 3	1,102	5,863	7,958	6,722

B.

− 3	5,349	4,177	4,132	6,554

Word Problem

C. Jeremy scooped 2,938 pumpkin seeds out of his pumpkin. Andrea scooped 39 more pumpkin seeds out of her pumpkin than Jeremy. Justice scooped 49 more pumpkin seeds out of her pumpkin than Andrea. How many pumpkin seeds did Andrea and Justice scoop out of their pumpkins?

Create your own math problem and explain your solution.

Math Problem:

Explanation/Solution:

Warm Up

Name: _____ **Date:** _____

Add.

A.
```
  1,043
+ 7,539
-------
```

B.
```
  6,904
+ 1,519
-------
```

C.
```
  3,768
+ 1,098
-------
```

D.
```
  6,419
+ 1,572
-------
```

E.
```
  4,177
+ 2,691
-------
```

Word Problem

F. Did more people attend the night performances or the afternoon performances?

Performance	Attendance
Friday Night	1,091
Saturday Afternoon	3,132
Saturday Night	3,191
Sunday Afternoon	1,071

Create your own math problem and explain your solution.

Math Problem:

Explanation/Solution:

Warm Up

49

Solve.

A.
$24.49
+ $38.91
_____

B.
$87.32
− $78.86
_____

C.
$48.86
− $41.32
_____

D.
$62.39
− $19.99
_____

E.
$16.11
+ $21.02
_____

Word Problem

F. Jackie had $10.61. She bought a pennant. How much change was she given? Does Jackie have enough money left to buy a cap?

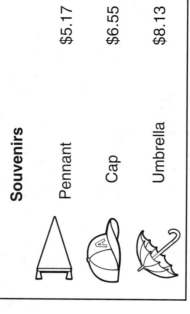

Souvenirs

Pennant	$5.17
Cap	$6.55
Umbrella	$8.13

Create your own math problem and explain your solution.

Math Problem:

Explanation/Solution:

Answers: A. $63.40 B. $8.46 C. $7.54 D. $42.40 E. $37.13 F. $5.44 in change; No

Warm Up

Name: _____ **Date:** _____

Write the numbers in order, smallest to largest.

A.

6,693 2,373 1,058

___ , ___ , ___

B.

9,867 1,745 2,101

___ , ___ , ___

C.

4,959 3,431 7,107

___ , ___ , ___

D.

5,857 4,436 5,794

___ , ___ , ___

Word Problem

E. The Big League Stadium seats 6,721 fans. On opening night, 9,774 fans attended. How many fans had to stand to watch the game?

Create your own math problem and explain your solution.

Math Problem:

Explanation/Solution:

Name: _____

Date: _____

Warm Up

Write the numbers that are ten less and ten more.

A. _____; 9,134; _____

B. _____; 4,413; _____

C. _____; 1,109; _____

D. _____; 8,857; _____

E. _____; 8,293; _____

F. _____; 4,625; _____

Word Problem

G. Gianni made 2,419 origami animals. Sylvia made ten more origami animals than Gianni. Bert made ten fewer origami animals than Gianni. How many origami animals did Sylvia and Bert make?

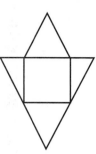

Create your own math problem and explain your solution.

Math Problem: _____

Explanation/Solution: _____

Answers: A. 9,124; 9,144 B. 4,403; 4,423 C. 1,099; 1,119 D. 8,847; 8,867 E. 8,283; 8,303 F. 4,615; 4,635 G. Sylvia 2,429; Bert 2,409

Name: _____ **Date:** _____

Warm Up

52

Rewrite the length of each shoreline in standard form. (Information from *Scholastic Book of World Records*. 2001. Page 109.)

A. Florida's shoreline is eight thousand, four hundred twenty-six miles.

_____

B. California's shoreline is three thousand, four hundred twenty-seven miles.

_____

C. Alaska's shoreline is thirty-three thousand, nine hundred four miles.

_____

D. Louisiana's shoreline is seven thousand, seven hundred twenty-one miles.

_____

Word Problem

E. If all of the shorelines were subtracted from Alaska's shoreline, how many miles would be left in the Alaskan shoreline?

Create your own math problem and explain your solution.

Math Problem:

Explanation/Solution:

_____

Name:

Date:

Warm Up

Rewrite in standard form.

A. 6,000 + 400 + 10 + 2

_____

B. 5,000 + 800 + 90 + 1

_____

C. 8,000 + 600 + 10 + 3

_____

D. 7,000 + 100 + 30 + 6

_____

Word Problem

Rewrite the math problem in standard form. Solve.

E. Last year, Louisa drove her taxi nine thousand, nine hundred sixteen miles. This year, Louisa drove her taxi two thousand, six hundred seventy fewer miles than last year. How many miles has Louisa driven this year?

Create your own math problem and explain your solution.

Math Problem:

Explanation/Solution:

_____

Name: _____ **Date:** _____

Warm Up

Rewrite each problem rounding each number to the nearest thousand. Add.

A.
2,576 _____
1,157 _____
1,021 _____
+ 9,888 _____
+ _____

B.
2,172 _____
1,144 _____
5,104 _____
+ 9,187 _____
+ _____

C.
4,672 _____
1,088 _____
9,647 _____
+ 6,139 _____
+ _____

D.
4,599 _____
4,466 _____
6,331 _____
+ 323 _____
+ _____

Word Problem

E. Write the math problem. Round each number to the nearest thousand.

Attendance Figures	
Glee Club	9,911
Jazz Festival	5,784
Concert Band	3,927
Harps	352
Country Music	4,686
Rock-n-Roll	1,045

1. Glee Club +
Jazz Festival +
Concert Band

_____
_____
+ _____
+ _____

2. Harps + Country
Music + Rock-n-
Roll

_____
_____
+ _____
+ _____

3. Glee Club +
Concert Band +
Country Music

_____
_____
+ _____
+ _____

4. Jazz Festival +
Harps + Rock-n-
Roll

_____
_____
+ _____
+ _____

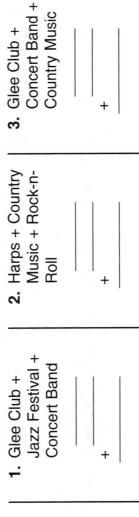

Create your own math problem and explain your solution.

Math Problem: _____

Explanation/Solution: _____

Name: _____

Date: _____

Warm Up

Subtract.

A.
```
  67,258
- 24,197
```
_____

B.
```
  67,592
- 39,710
```
_____

C.
```
  52,310
- 10,110
```
_____

D.
```
  96,886
- 87,282
```
_____

E.
```
  93,612
- 56,324
```
_____

Word Problem

F. On opening night, 47,534 people attended the premiere of *Lizard vs. Iguana Man.* The following night 43,099 people attended the movie. How many more people saw the movie on its opening night?

Create your own math problem and explain your solution.

Math Problem:

Explanation/Solution:

_____

Answers: A. 43,061 B. 27,882 C. 42,200 D. 9,604 E. 37,288 F. 4,435 more people

56

Warm Up

Write the name of the month for each holiday or special occasion.

A. Valentine's Day _____

B. St. Patrick's Day _____

C. Thanksgiving _____

D. Mother's Day _____

E. Ground Hog's Day _____

Word Problem

F. Use a calendar (either a current one or an out-of-date one) to answer each question.

1. How many days are there between Ground Hog's Day and Valentine's Day? _____

2. How many days are there between Halloween and Thanksgiving? _____

3. Which holiday will be next—Mother's Day or Father's Day? _____

4. How many more days until your birthday? _____

5. What is the name of the next season? _____

6. How many Fridays are there from now to the end of the year? _____

Create your own math problem and explain your solution.

Math Problem:

Explanation/Solution:

_____

Answers: A. February B. March C. November D. May E. February F1. 12 days F2–6. Answers will vary.

Name: _____ **Date:** _____

Warm Up

Write the time on the line.

A. _____ B. _____ C. _____ D. _____ E. _____

Word Problem

F. Complete the train schedule. What time does Train D depart?

Never-Late Commuter Train
A train leaves every 5 minutes!

Train A   9:25
Train B   _____
Train C   _____
Train D   _____

Create your own math problem and explain your solution.

Math Problem:

Explanation/Solution: _____

Answers: A. 4:40 B. 2:50 C. 1:10 D. 5:35 E. 6:20 F. Schedule: 9:30, 9:35, 9:40, Train D will depart at 9:40.

Warm Up

Draw the hands on the clock to show the time.

A.	B.	C.	D.	E.

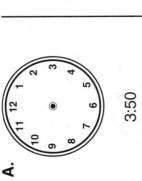

A. 3:50

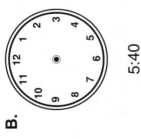

B. 5:40

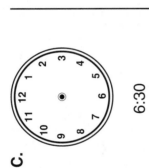

C. 6:30

D. 4:05

E. 8:10

Word Problem

Dinner in a Jiffy!

- Stir the ingredients for 6 minutes.
- Pour in a pan and for 2 minutes evenly distribute the mixture.
- Put in the oven and bake for 23 minutes.
- Let cool for 8 minutes.

F. How long will it take to prepare this meal and serve it to the family?

Create your own math problem and explain your solution.

Math Problem: _____

Explanation/Solution: _____

Answers: *A. hour hand between the 3 and 4, minute hand on the 10 B. hour hand between the 5 and 6, minute hand on the 8 C. hour hand between the 6 and 7, minute hand on the 6 D. hour hand on the 4, minute hand on the 1 E. hour hand on the 8, minute hand on the 2 F. 39 minutes*

Name: _____

Date: _____

Warm Up

Answer each question.

A. How many months in one year?

_____

B. Which month has 28 days?

_____

C. Which months have 30 days?

_____
_____
_____
_____

D. Which months have 31 days?

_____
_____
_____
_____

E. What are the names of the four seasons?

_____
_____
_____
_____

Word Problem

F. Use a calendar for this month to answer the questions.

1. What is the name of this month? _____

2. What is the date for the third Tuesday of this month? _____

3. On what day does the month begin? _____

4. What is the day and date for the last day of this month? _____

5. What is the date for the last Friday of this month? _____

6. Are there any holidays for this month? If yes, name them. _____

Create your own math problem and explain your solution.

Math Problem: _____

Explanation/Solution: _____

Answers: A. 12 B. February C. April, June, September, November D. January, March, May, July, August, October, December E. winter, spring, summer, fall (or autumn)

F1.–6. Answers will vary.

Warm Up

68

Write the ratio.

A. 3 parts water to 2 parts powdered drink

_____

B. 1 cup of rice to 2 cups of water

_____

C. 2 tablespoons of butter to 1 tablespoon of honey

_____

D. 1 gallon of white paint to 2 gallons of blue paint

_____

Word Problem

E. For every bad apple there are 4 good apples. How many bad apples would you need to go through in order to find 20 good apples? Make a table.

Bad Apples	1				
Good Apples	4				

Create your own math problem and explain your solution.

Math Problem:

Explanation/Solution:

_____

Answers: A. 3:2 B. 1:2 C. 2:1 D. 1:2 E. bad apples: 1, 2, 3, 4, 5; good apples: 4, 8, 12, 16, 20; You would need to go through 5 bad apples to find 20 good apples.

61

Write the wins and losses for each year using the graph below. (■ = wins   ▓ = losses)

A. 1st Year

Wins: _____

Losses: _____

B. 2nd Year

Wins: _____

Losses: _____

C. 3rd Year

Wins: _____

Losses: _____

D. 4th Year

Wins: _____

Losses: _____

Word Problem

1. How many games were played each year? _____
2. What year did the Blue Sox win the most games? _____
3. In which year did they lose the most games? _____
4. During which years did the Blue Sox win more games than they lost? _____
5. Which year did they have the same number of wins and losses? _____

E.

Create your own math problem and explain your solution.

Math Problem:

Explanation/Solution:

62

Name: _____

Date: _____

Warm Up

Draw the picture. Write the multiplication problem.

A. 5 rows of stars with 4 stars in each row

_____

B. 4 rows of stars with 4 stars in each row

_____

C. 2 rows of stars with 6 stars in each row

_____

D. 3 rows of stars with 5 stars in each row

_____

Word Problem

E. Cheryl made 7 stars in one minute. How many stars could Cheryl make in 5 minutes?

Create your own math problem and explain your solution.

Math Problem:

Explanation/Solution:

Answers: A. 5 x 4 = 20 B. 4 x 4 = 16 C. 2 x 6 = 12 D. 3 x 5 = 15 E. 35 stars

Warm Up

Name: _____ **Date:** _____

Write the multiplication problem.

A. _____

B. _____

C. _____

D. _____

Word Problem

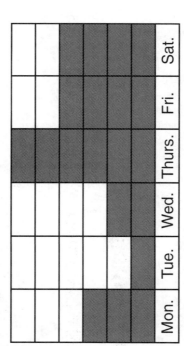

▨ = 5 arrows

	Mon.	Tue.	Wed.	Thurs.	Fri.	Sat.

E.

1. How many arrows were sold on Tuesday? _____

2. On which two days were the same number of arrows sold? _____

3. Were there more arrows sold on Monday and Tuesday or on Wednesday and Thursday? _____

Create your own math problem and explain your solution.

Math Problem:

Explanation/Solution:

Answers: A. 3 x 5 = 15 B. 4 x 5 = 20 C. 1 x 5 = 5 D. 2 x 5 = 10 E1. 5 arrows E2. Fri. and Sat. E3. Wed. and Thurs.

64

Name: _____

Date: _____

Warm Up

Multiply.

A.

$3 \times 5 =$ _____

B.

$5 \times 3 =$ _____

C.

$4 \times 6 =$ _____

D.

$6 \times 4 =$ _____

E.

$2 \times 8 =$ _____

F.

$8 \times 2 =$ _____

G.

$1 \times 10 =$ _____

H.

$10 \times 1 =$ _____

I.

$0 \times 9 =$ _____

J.

$9 \times 0 =$ _____

Word Problem

K. There are three rolls of paper towels in each pack. How many rolls of paper towels are there in four packs?

Create your own math problem and explain your solution.

Math Problem:

Explanation/Solution:

Answers: A. 1, 2, 4, 8 B. 1, 2, 3, 4, 6, 12 C. 1, 3, 9 D. 1, 2, 3, 6 E. Small (1 x 8 = 8), Medium (3 x 8 = 24), Large (2 x 8 = 16)

65

Name: _____ **Date:** _____

Warm Up

Write the factors for each number.

A. 8 _____

B. 12 _____

C. 9 _____

D. 9 _____

Word Problem

E. We Make Wheels carries wheels for roller skates in three different sizes—small, medium, and large. Each set contains 8 wheels. How many wheels are there in each size?

= 1 set of 8 wheels

	Small	Medium	Large

Create your own math problem and explain your solution.

Math Problem:

_____

Explanation/Solution:

Name: _____

Date: _____

Warm Up

A. Complete the table.

x	0	1	2	3	4	5	6
7							

B. Write the factors for seven.

_____

C. Circle the multiples of 7.

8 15 21 9 56

7 2 10 14 70

Word Problem

D. Write the multiplication problem. Solve.

Little Jane was at the archery practice range. She hit the following numbers: 3, 5, 5, 10, 5, 3, 7.

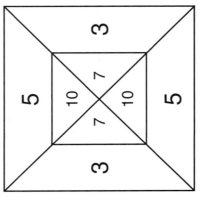

Create your own math problem and explain your solution.

Math Problem:

_____

Explanation/Solution:

67

Warm Up

Write the multiplication problem.

A. number of legs on 6 ladybugs

_____

B. number of wings on 5 butterflies

_____

C. number of legs on 10 dogs

_____

D. number of tails on 12 cats

_____

Word Problem

E. There were 9 ladybugs sitting on a flower. Four of the ladybugs flew away. Two more ladybugs came and landed on the flower. How many legs are on the flower?

Create your own math problem and explain your solution.

Math Problem: _____

Explanation/Solution: _____

Answers: A. 6 x 6 = 36 B. 2 x 5 = 10 C. 4 x 10 = 40 D. 1 x 12 = 12 E. 9 – 4 + 2 = 7 x 6 (legs) = 42 legs are on the flower

Warm Up

A. Write the rule for multiplying a number by 9.

_____

_____

B. Write the missing factor.

9 x _____ = 36

C. Complete the number pattern.

9, _____, 27, _____,

_____, _____, _____,

_____, _____, _____

D. Solve.

9 x (3 x 2) = _____

Word Problem

E. The change machine charges 9¢ for every dollar. How much would it cost to change $10 into coins?

Create your own math problem and explain your solution.

Math Problem:

Explanation/Solution:

Answers: A. _The digits in the product added together always equal 9._ B. _4_ C. _(9), 18, (27), 36, 45, 54, 63, 72, 81, 90_ D. _9 x (3 x 2) = 9 x 6 = 54_ E. _9¢ x $10.00 = 90¢_

68

Warm Up

Write the factors for each number.

A. 20 _____

B. 12 _____

C. 13 _____

D. 25 _____

E. 19 _____

Word Problem

F. Write the factors for each number. Circle the common factors used in both numbers.

24: _____

40: _____

Create your own math problem and explain your solution.

Math Problem:

Explanation/Solution: _____

Answers: A. 1, 2, 4, 5, 10, 20 B. 1, 2, 3, 4, 6, 12 C. 1, 13 D. 1, 5, 25 E. 1, 19 F. 24: 1, 2, 3, 4, 6, 8, 12, 24; 40: 1, 2, 4, 5, 8, 10, 20, 40; common factors: 1, 2, 4, 8

70

Name: _____ Date: _____

Warm Up

Divide.

A.

$3\overline{)9}$

B.

$2\overline{)8}$

C.

$2\overline{)10}$

D.

$4\overline{)8}$

E.

$4\overline{)12}$

Word Problem

F. Matt baked a dozen cookies. He shared them equally with 3 of his friends. How many cookies did Matt and each friend receive?

Create your own math problem and explain your solution.

Math Problem:

Explanation/Solution:

Warm Up

Name: _____

Date: _____

Divide.

A.

$4\overline{)20}$

_____

B.

$4\overline{)36}$

_____

C.

$4\overline{)80}$

_____

D.

$4\overline{)4}$

_____

E.

$4\overline{)40}$

_____

Word Problem

F. At the soccer playoffs there were 56 kids on 4 teams. How many kids were on each team?

Create your own math problem and explain your solution.

Math Problem:

Explanation/Solution: _____

72

Name: _____

Date: _____

Warm Up

Divide.

A.

$5\overline{)5}$

_____

B.

$5\overline{)25}$

_____

C.

$5\overline{)10}$

_____

D.

$5\overline{)100}$

_____

E.

$5\overline{)45}$

_____

Word Problem

F. During weight lifting the students are able to lift 1/5 of their body weight. If a student weighs 50 pounds, how much weight should the student lift?

Create your own math problem and explain your solution.

Math Problem:

Explanation/Solution: _____

73

Name:

Date:

Complete each pattern.

A.

3, 11, 6, 14, 9, ____, ____

B.

9, 14, 11, 16, 13, ____, ____

C.

6, 2, 9, 5, 12, ____, ____

D.

2, 6, 5, 9, 8, ____, ____

Word Problem

E. Complete the table to find the pattern.

Triplets	3		12		
Hands	6			36	

Write the division problem that tells the number of triplets if there were 20 hands.
Write the multiplication problem that tells the number of hands if there were 30 triplets.

Create your own math problem and explain your solution.

Math Problem: _____

Explanation/Solution:

Answers: A. 17, 12 B. 18, 15 C. 8, 15 D. 12, 11 E. triplets: (3), 6, 9, (12), 15, 18, 21, 24; hands: (6), 12, 18, 24, 30 (36), 42, 48; division problem: 20 ÷ 2 = 10 triplets; multiplication problem: 30 x 2 = 60 hands

74

Warm Up

Subtract.

A.

$$\frac{6}{7} - \frac{5}{7} = \text{____}$$

B.

$$\frac{18}{20} - \frac{10}{20} = \text{____}$$

C.

$$\frac{12}{15} - \frac{6}{15} = \text{____}$$

D.

$$\frac{11}{18} - \frac{5}{18} = \text{____}$$

E.

$$\frac{3}{17} - \frac{2}{17} = \text{____}$$

Word Problem

F. Find the number of gumballs each person bought. The Really Giant Gumball Machine holds 24 giant gumballs.

1. Mabel bought 1/4 of the gumballs. How many did she buy? _____

2. Willie bought 1/8 of the gumballs. How many did he buy? _____

3. Shirley bought twice as many gumballs as Mabel. How many did she buy? _____
   What is the fractional amount? _____

4. How many gumballs are left? _____ What is the fractional amount? _____

Create your own math problem and explain your solution.

Math Problem:

Explanation/Solution: _____

Answers: A. 1/7 B. 8/20 C. 6/15 D. 6/18 E. 1/17 F1. 6 gumballs F2. 3 gumballs F3. 12 gumballs,1/2 F4. 3 gumballs are left, 1/8

Name: _____ **Date:** _____

Warm Up

Complete the table.

A.

÷	5	10	15	20	25	30
5						

B.

÷	6	12	18	24	30	36
6						

Word Problem

C. Desiree had two dozen eggs. She put the eggs into 6 different groups. How many eggs are in each group?

Create your own math problem and explain your solution.

Math Problem:

Explanation/Solution:

Answers: A. 1, 2, 3, 4, 5, 6 B. 1, 2, 3, 4, 5, 6 C. 24 ÷ 6 = 4 eggs in each group

78

Warm Up

Name: _____

Date: _____

Rewrite each problem. Solve.

A. _____

$49 \div 7 =$

B. _____

$14 \div 7 =$

C. _____

$28 \div 7 =$

D. _____

$7 \div 7 =$

E. _____

$35 \div 7 =$

Word Problem

F. There are 28 days in February. How many times does each day of the week occur in February?

Create your own math problem and explain your solution.

Math Problem:

Explanation/Solution: _____

Answers: A. $7\overline{)49} = 7$ B. $7\overline{)14} = 2$ C. $7\overline{)28} = 4$ D. $7\overline{)7} = 1$ E. $7\overline{)35} = 5$ F. $7\overline{)28} = 4$

Name:

Date:

Warm Up

Circle the answer.

A. Which number can be evenly divided by 8?

46   64

B. Which number can be evenly divided by 7?

56   65

C. Which number can be evenly divided by 6?

54   45

D. Which number can be evenly divided by 9?

66   63

Word Problem

E. Using the chart, read the clues to discover the mystery number.

- The number is divisible by 7.

- It is an even number.

- It is a multiple of 4.

What is the mystery number? _____

2	3	4	5	6	7	8	9
12	13	14	15	16	17	18	19
22	23	24	25	26	27	28	29
32	33	34	35	36	37	38	39
42	43	44	45	46	47	48	49

Create your own math problem and explain your solution.

Math Problem:

Explanation/Solution:

_____

78

Name: _____ Date: _____

Warm Up

Divide by 10 to convert each measurement from millimeters (mm) to centimeters (cm).

A.

10 mm

_____ cm

B.

70 mm

_____ cm

C.

80 mm

_____ cm

D.

90 mm

_____ cm

E.

100 mm

_____ cm

Word Problem

F. Latasha, Julia, Jenica, and Hector made paper airplanes. Each person threw his or her airplane and measured the distance it flew.

Paper Airplanes	
Latasha	350 mm
Julia	31 cm
Jenica	49 cm
Hector	510 mm

1. Whose airplane flew the shortest distance? _____

2. Whose airplane flew more than 45 cm? _____

3. Whose airplanes flew longer distances—Latasha's and Julia's or Hector's and Jenica's? _____

Create your own math problem and explain your solution.

Math Problem:

Explanation/Solution:

_____

Answers: A. 1 cm B. 7 cm C. 8 cm D. 9 cm E. 10 cm F1. Julia's F2. Jenica's and Hector's F3. Hector's and Jenica's

79

Name: _____ **Date:** _____

Use >, <, or = symbols to compare each set of measurements.

A.

33 mm ◯ 3 cm

B.

68 mm ◯ 5 cm

C.

11 mm ◯ 4 cm

D.

60 mm ◯ 6 cm

E.

79 mm ◯ 1 cm

Word Problem

F. Antony Ant went for a walk. He climbed over the tall mountains, walked under the flying dragon, and climbed over the stinky shoe. How far did Antony Ant walk in all? Write the answer in millimeters and centimeters.

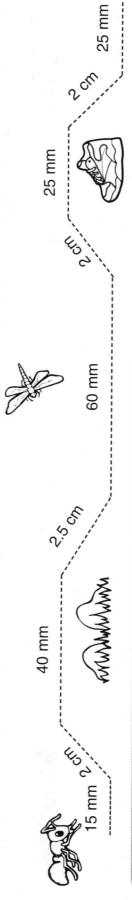

15 mm 2 cm

40 mm 2.5 cm

60 mm 2 cm

25 mm 2 cm 25 mm

Create your own math problem and explain your solution.

Math Problem:

Explanation/Solution:

Answers: A > B > C < D = E > F. 250 mm or 25 cm

Warm Up

Find the number of centimeters (cm) and millimeters (mm) for each number of meters (m). Remember the following formulas:
cm = m x 100 and mm = m x 1,000.

A. 6 meters

_____ cm

_____ mm

B. 5 meters

_____ cm

_____ mm

C. 1 meter

_____ cm

_____ mm

D. 4 meters

_____ cm

_____ mm

E. 7 meters

_____ cm

_____ mm

Word Problem

F. Make a list of items that could be measured in millimeters, centimeters, and meters.

Millimeters

1. _____

2. _____

3. _____

4. _____

5. _____

6. _____

7. _____

8. _____

Centimeters

1. _____

2. _____

3. _____

4. _____

5. _____

6. _____

7. _____

8. _____

Meters

1. _____

2. _____

3. _____

4. _____

5. _____

6. _____

7. _____

8. _____

Create your own math problem and explain your solution.

Math Problem:

Explanation/Solution: _____

Answers: A. 600 cm; 6,000 mm B. 500 cm; 5,000 mm C. 100 cm; 1,000 mm D. 400 cm; 4,000 mm E. 700 cm; 7,000 mm F. Answers will vary.

Warm Up

Rank the following measurements in order, smallest to greatest.

A. 9 cm   1 mm
8 m   10 cm

B. 81 cm   3 mm
600 cm   2 m

C. 7 m   98 mm
12 cm   833 mm

D. 101 cm   1 m
27 cm   73 mm

___, ___, ___, ___

___, ___, ___, ___

___, ___, ___, ___

___, ___, ___, ___

Word Problem

E. Edwina, Eugenia, and Edgar were having a jumping contest. Edwina jumped 310 cm. Eugenia jumped 3,965 mm. Edgar jumped 3 meters. Who jumped the farthest distance?

Create your own math problem and explain your solution.

Math Problem:

_____

Explanation/Solution:

_____

Warm Up

Name: _____

Date: _____

82

Write the problem. Multiply.

A. What is the value of five nickels?

_____

B. What is the value of ten pennies?

_____

C. What is the value of nine dimes?

_____

D. What is the value of five one-dollar bills?

_____

Word Problem

E. Shane earns 6¢ for every magazine subscription he sells. Shane has sold 7 subscriptions. How much money has he earned?

Create your own math problem and explain your solution.

Math Problem:

Explanation/Solution:

_____

Name: _____

Date: _____

Warm Up

Round each number.

A. Round to the nearest $1.00.

$8.42

_____

B. Round to the nearest $10.00.

$63.10

_____

C. Round to the nearest $100.00.

$101.42

_____

D. Round to the nearest $1,000.

$8,124.96

_____

E. Round to the nearest $10,000.

$18,249.16

_____

Word Problem

F. At Joe's Diner, 1,041,101 hamburgers were sold.

	Cost	Sold	Profit
1 hamburger	$0.30	$0.90	_____

1. Find the profit for one hamburger. _____

2. Find the total cost for making 1,041,101 hamburgers. _____

3. Find the total profit (selling price – cost) for 1,041,101 hamburgers. _____

Create your own math problem and explain your solution.

Math Problem: _____

Explanation/Solution: _____

Name: _____ **Date:** _____

Warm Up

Use the price chart below to solve each problem.

A. 3 boxes of chalk

_____

B. 4 tubes of oil paint

_____

C. 9 drawing pads

_____

D. 9 easels

_____

E. 6 small brushes

_____

Word Problem

F. Joselyn wants to buy 1 small brush, 2 easels, and 4 drawing pads. Does she have enough money?

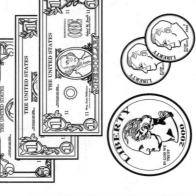

Art Supplies	
Box of Chalk	$0.82
Small Brush	$2.88
Easel	$4.89
Drawing Pad	$0.93
Oil Paint	$0.71

Create your own math problem and explain your solution.

Math Problem: _____

Explanation/Solution: _____

Answers: A. $2.46 B. $2.84 C. $8.37 D. $44.01 E. $17.28 F. No (The items would cost $16.38.)

85

Warm Up

Write the time two different ways.

A.
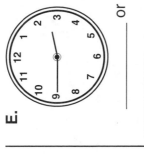
_____ or _____

B.
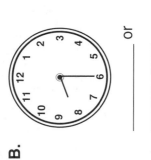
_____ or _____

C.
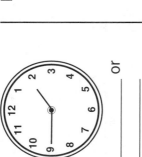
_____ or _____

D.
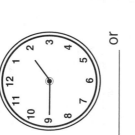
_____ or _____

E.

_____ or _____

Word Problem

F. The Express Line Subway Car takes 15 minutes from one stop to the next. If the Express Line Subway Car left Station #1 at 8:45, what time would it reach Station #3? What is the elapsed time?

Create your own math problem and explain your solution.

Math Problem:

Explanation/Solution:
_____

Answers: A. 6:15 or quarter past 6 B. 8:30 or half past 8 C. 1:45 or a quarter till 2 D. 5:15 or a quarter past 5 E. 2:45 or a quarter till 3 F. It would reach Station #3 at 9:15. The elapsed time is 30 minutes.

88

Warm Up

Rewrite each time in hours and minutes.

A. 68 minutes

____ hour
____ minutes

B. 85 minutes

____ hour
____ minutes

C. 93 minutes

____ hour
____ minutes

D. 74 minutes

____ hour
____ minutes

E. 61 minutes

____ hour
____ minutes

Word Problem

F. Elisabeth spent 58 minutes practicing the violin, 13 minutes practicing the piano, and 72 minutes practicing the flute. How many minutes did Elisabeth practice in all? Write the time in standard form.

Create your own math problem and explain your solution.

Math Problem:

Explanation/Solution: _____

Answers: A. 1 hour 8 minutes B. 1 hour 25 minutes C. 1 hour 33 minutes D. 1 hour 14 minutes E. 1 hour 1 minute F. 143 minutes, 2 hours 23 minutes

87

Warm Up

Use >, <, or = to compare each amount of time.

A.

62 min. ◯ 2 hrs.

_____

B.

1/2 hr. ◯ 16 min.

_____

C.

83 min. ◯ 1 hr.

_____

D.

103 min. ◯ 1 ½ hrs.

_____

Word Problem

E. There are two hours left until the World's Greatest Theme Park closes for the night. Figure out the greatest number of rides that can be taken before the park closes. Will there be any time left over?

Wait Time	
Thunder Roller Coaster	54 min.
Scary Ride	20 min.
Merry-Go-Round	71 min.
Wet and Wild Log Ride	83 min.
Train Ride	5 min.

Create your own math problem and explain your solution.

Math Problem:

_____

Explanation/Solution:

Answers: A. < B. > C. > D. > E. 3 rides, yes there will be time left over.

Name: _____

Date: _____

Warm Up

Write the time.

A. A cake was put in the oven at 3:07 and baked for 34 minutes. What time was the cake done?

_____

B. Cookies were put on the cookie sheet at 9:16 and baked for 6 minutes. What time were the cookies done?

_____

C. The turkey was roasted for 1 hour and 5 minutes and was taken out of the oven at 7:50. What time was the turkey put into the oven?

_____

Word Problem

D. Each custard dessert takes 7 minutes to prepare. How long would it take to make 14 custard desserts?

Create your own math problem and explain your solution.

Math Problem:

Explanation/Solution:

_____

68

Warm Up

Write the fraction for the shaded part of the shape.

A. _____

B. _____

C. _____

D. _____

E. _____

Word Problem

F. What information would you need to solve the word problem?

Which cost more: $14.00 for a pie or $1.50 for each slice of an apple pie?

Answers: A. 5/9 B. 2/6 or 1/3 C. 1/7 D. 3/8 E. 11/12 F. How many slices are in the apple pie?

Create your own math problem and explain your solution.

Math Problem:

Explanation/Solution: _____

Warm Up

90

Write the fraction for the shaded part of the shape.

A.   _____

B.   _____

C.   _____

D.   _____

E.   _____

Word Problem

F.

Mom's Restaurant	
Pie	$1.79 a slice
With ice cream	add $1.93

1. Each slice of an 8-slice pie costs $1.79. How much for the whole pie? _____

2. If the slice were served with ice cream, what would the total cost be for one slice of pie? _____

Create your own math problem and explain your solution.

Math Problem:

Explanation/Solution:

_____

91

Warm Up

Divide each shape. Shade to show the fraction.

A.
[ ]

$\frac{1}{2}$

B.
[ ]

$\frac{3}{4}$

C.
[ ]

$\frac{2}{5}$

D.
[ ]

$\frac{2}{9}$

E.
[ ]

$\frac{3}{8}$

Word Problem

F. Sixteen students were asked, "What do you like to do when you have some free time?" Answer each question using the information from the pie chart.

cards

phone

board games

music

1. Do more students like to play cards or listen to music? _____

2. Do fewer students like to play board games or play cards? _____

3. How many students like to talk on the phone? _____

4. How many students like to play cards? _____

5. How many students like to listen to music? _____

Create your own math problem and explain your solution.

Math Problem:

Explanation/Solution:

_____

Answers: A.–E. Students need to divide each rectangle into the correct number of parts and color the appropriate sections. F1. music F2. board games F3. 2 F4. 4 F5. 8

92

Name: _____ **Date:** _____

Warm Up

Divide each shape. Shade to show the fraction.

A. _____

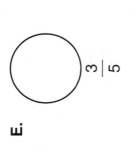

$\frac{1}{4}$

B. _____

$\frac{2}{6}$

C. _____

$\frac{3}{9}$

D. _____

$\frac{5}{8}$

E. _____

$\frac{3}{5}$

Word Problem

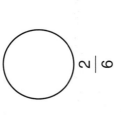

funny | scary | action

F. Four kids were asked to name their favorite kind of movie. Use the pie chart to answer each question.

1. What fraction of kids like action movies the best? _____

2. What fraction of kids like funny movies the best? _____

3. Did more kids like action movies or scary movies? _____

4. How many kids like action movies? _____

Create your own math problem and explain your solution.

Math Problem:

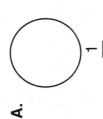

Explanation/Solution: _____

Answers: A.–E. Students should divide each shape and color to match the fractions. F1. 1/2 F2. 1/4 F3. action movies F4. 2 kids

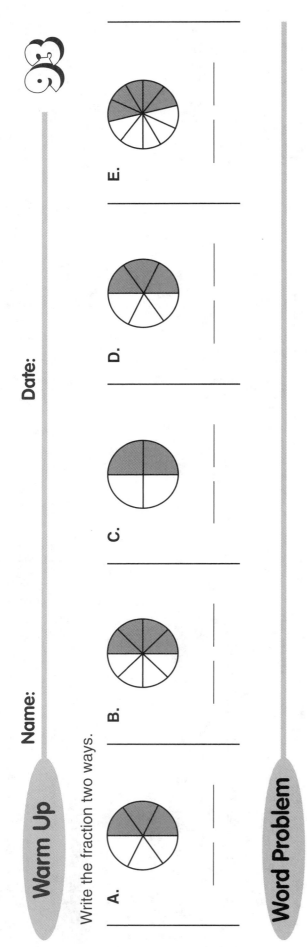

Name: _____

Date: _____

ℬℬ

Warm Up

Write the fraction two ways.

A. _____ _____

B. _____ _____

C. _____ _____

D. _____ _____

E. _____ _____

Word Problem

F. Percy sold half of his raffle tickets. Sharla sold 6/12 of her raffle tickets. Who sold more tickets?

Create your own math problem and explain your solution.

Math Problem:

Explanation/Solution: _____

Answers: A. 3/6, 1/2 B. 4/8, 1/2 C. 2/4, 1/2 D. 3/6, 1/2 E. 5/10, 1/2 F. Both of them sold the same number of tickets!

Name: _____

Date: _____

Warm Up

94

Simplify each fraction.

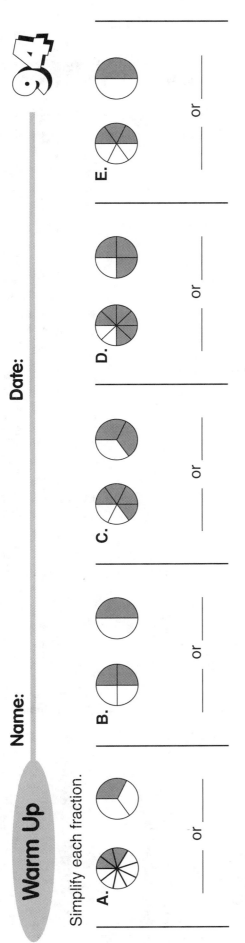

A. _____ or _____

B. _____ or _____

C. _____ or _____

D. _____ or _____

E. _____ or _____

Word Problem

F. Jacob's mom made a delicious chocolate pie. Jacob asked for the biggest piece of the pie. His mother said, "Your choice is 1/3 of the pie or 3/5 of the pie." Which choice is the larger piece?

Create your own math problem and explain your solution.

Math Problem:

Explanation/Solution:

Answers: A. 3/9 or 1/3 B. 2/4 or 1/2 C. 4/6 or 2/3 D. 6/8 or 3/4 E. 3/6 or 1/2 F. 3/5 is the larger piece of pie.

Name: _____ **Date:** _____

Warm Up

Simplify each fraction.

A.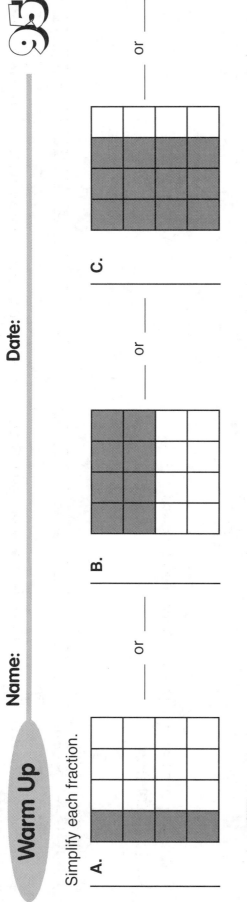

_____ or _____

B.

_____ or _____

C.

_____ or _____

Word Problem

D. Devin's photo album can hold one dozen pictures. He has already filled 1/3 of the photo album. How many pictures does Devin have in his album?

Create your own math problem and explain your solution.

Math Problem:

Explanation/Solution: _____

Name: _____

Date: _____

Warm Up

Shade 1/2 of each shape. Write the fraction two ways.

A.

_____ and _____

B.

_____ and _____

C.

_____ and _____

D.

_____ and _____

E.

_____ and _____

Word Problem

F. Identify the missing piece of information for each question.

1. Asia ate 1/2 of the blueberry pie. How many slices did Asia eat? _____

2. If there were 8 slices of pie, how much did each slice cost? _____

3. China ate 3 slices of pie. How many slices are left? _____

4. Aleman made 10 delicious pies for the bake sale. He sold almost all of the pies. How many pies does Aleman have left? _____

Create your own math problem and explain your solution.

Math Problem:

Explanation/Solution:

Answers: A. 5/10 and 1/2 B. 3/6 and 1/2 C. 4/8 and 1/2 D. 2/4 and 1/2 E. 6/12 and 1/2 F1. How many slices were there to begin with? F2. What was the total cost for one pie? F3. How many slices were there to begin with? F4. How many pies did Aleman sell?

Name: _____ **Date:** _____

Warm Up

Complete each table.

A.

	+ 2	− 2	× 2	÷ 2
12				

B.

	+ 3	− 3	× 3	÷ 3
15				

C.

	+ 4	− 4	× 4	÷ 4
20				

D.

	+ 5	− 5	× 5	÷ 5
30				

Word Problem

E. Elliot needs to buy 5 pencils. Which store has the best pencil price?

We Sell Pencils
10 pencils for $3.00

Pencils-R-Us
3 pencils for $1.11

Bargain Pencils
$0.40 a pencil

Create your own math problem and explain your solution.

Math Problem:

_____

Explanation/Solution:

_____

Answers: A. 14, 10, 24, 6 B. 18, 12, 45, 5 C. 24, 16, 80, 5 D. 35, 25, 150, 6 E. We Sell Pencils has the best price.

Warm Up

😎

Write the missing number.

A.

$24 \div \underline{\hspace{1cm}} = 4$

B.

$4 \times 6 = \underline{\hspace{1cm}}$

C.

$\underline{\hspace{1cm}} \div 10 = 7$

D.

$3 \times 6 = \underline{\hspace{1cm}}$

E.

$\underline{\hspace{1cm}} \div 9 = 9$

F.

$10 \times \underline{\hspace{1cm}} = 0$

G.

$10 \times \underline{\hspace{1cm}} = 20$

H.

$72 \div \underline{\hspace{1cm}} = 8$

I.

$4 \times \underline{\hspace{1cm}} = 32$

J.

$75 \div \underline{\hspace{1cm}} = 15$

Word Problem

K. Sally had 18 thimbles in her collection. Wally had 3 times as many thimbles in his collection. Maria had 1/2 the number of thimbles than Sally in her collection. How many thimbles do Wally and Maria each have?

Create your own math problem and explain your solution.

Math Problem:

Explanation/Solution:

Name: _____ **Date:** _____

Warm Up **99**

Write the two multiplication and two division problems for each set of numbers.

A. 2, 9, 18

___ x ___ = ___
___ x ___ = ___
___ ÷ ___ = ___
___ ÷ ___ = ___

B. 6, 7, 42

___ x ___ = ___
___ x ___ = ___
___ ÷ ___ = ___
___ ÷ ___ = ___

C. 8, 3, 24

___ x ___ = ___
___ x ___ = ___
___ ÷ ___ = ___
___ ÷ ___ = ___

D. 7, 5, 35

___ x ___ = ___
___ x ___ = ___
___ ÷ ___ = ___
___ ÷ ___ = ___

Word Problem

E.

1. There are 6 pallets. Each pallet holds 75 cans of soda. How many cans are there in all? _____

2. Each shelf in the store can hold 1/9 of the total number of soda cans. How many sodas can each shelf hold? _____

Create your own math problem and explain your solution.

Math Problem:

_____

Explanation/Solution:

Answers: A. 2 x 9 = 18, 9 x 2 = 18, 18 ÷ 2 = 9, 18 ÷ 9 = 2 B. 6 x 7 = 42, 7 x 6 = 42, 42 ÷ 7 = 6, 42 ÷ 6 = 7 C. 8 x 3 = 24, 3 x 8 = 24, 24 ÷ 8 = 3, 24 ÷ 3 = 8 D. 7 x 5 = 35, 5 x 7 = 35, 35 ÷ 7 = 5, 35 ÷ 5 = 7 E1. 450 cans E2. 50 cans

100

Name: _____

Date: _____

Warm Up

Use the >, <, or = to compare each set of numbers.

A.

4 x 10 ◯ 5 x 9

B.

7 x 9 ◯ 6 x 6

C.

4 x 8 ◯ 3 x 3

D.

6 x 5 ◯ 3 x 10

Word Problem

E. Garcelle earns 5¢ for every earthworm she collects. If it takes 20 worms to fill a can, how many would she need to fill in order to earn to $25.00?

Create your own math problem and explain your solution.

Math Problem:

Explanation/Solution:

Name: _____

Date: _____

Warm Up

Solve each problem.

A. There are 72 spider legs. How many spiders are there?

_____

B. There are 64 octopus arms. How many octopuses are there?

_____

C. There are 32 sides on the octagons. How many octagons are there?

_____

D. There are 16 cards with the number eight. How many sets of cards are there?

_____

Word Problem

E. Karen can run one mile in eight minutes. How many miles can Karen run in 40 minutes?

Create your own math problem and explain your solution.

Math Problem:

Explanation/Solution: _____

Answers: A. 9 spiders B. 8 octopuses C. 4 octagons D. 4 sets of cards E. 5 miles

Warm Up

Name: _____ **Date:** _____

Multiply.

A.
10 x 7 = _____

B.
7 x 10 = _____

C.
10 x 8 = _____

D.
8 x 10 = _____

E.
10 x 4 = _____

F.
4 x 10 = _____

G.
10 x 2 = _____

H.
2 x 10 = _____

Word Problem

I. Answer the problems below.

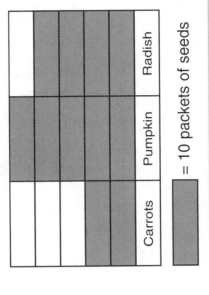

Carrots	Pumpkin	Radish

■ = 10 packets of seeds

1. Write the multiplication problem that tells how many packets of pumpkin seeds were sold. _____

2. If there are 20 carrot seeds in one packet, how many carrot seeds were sold in all? _____

3. If there were twice as many packets of radish seeds sold last month, how many packets would that be? _____

Create your own math problem and explain your solution.

Math Problem: _____

Answers: A. 70 B. 70 C. 80 D. 80 E. 40 F. 40 G. 20 H. 20 I1. 5 x 10 = 50 I2. 20 x 20 = 400 carrot seeds I3. 2 x 40 = 80 packets

Explanation/Solution: _____

Name: _____

Date: _____

Warm Up

Solve the problems.

A. Nine students can ride in each van. How many vans are needed for 27 students?

_____

B. Jason bowled 90 points in 10 games. What was the average number of points Jason bowled in each game?

_____

C. Complete the table.

÷	9	18	27	36	45	54	63
9							

Word Problem

D. Students go to school for 180 days for 9 months. As an average, how many days of school are there in each month?

Create your own math problem and explain your solution.

Math Problem:

Explanation/Solution: _____

104

Name: _____

Date: _____

Warm Up

A. Complete the table.

÷	10	20	30	40	50	60	70
10							

B. Write the rule for dividing a number by ten.

_____
_____
_____

Word Problem

C. The bag of chips has 1,110 calories for ten servings. How many calories are in one serving of chips?

Create your own math problem and explain your solution.

Math Problem:

_____

Explanation/Solution:

_____

Answers: A. 1, 2, 3, 4, 5, 6, 7 B. Sample answer: Move the decimal point in the dividend to the left the same number of times as the number of zeros in the divisor. The remaining number in the dividend is the answer. C. 111 calories per serving.

Name: _____ Date: _____

Warm Up

Use the >, <, or = symbol to compare each set of measurements. Remember, 3 teaspoons (tsp.) = 1 tablespoon (T).

A.

1 tsp. ◯ 2 T.

B.

6 tsp. ◯ 2 T.

C.

4 tsp. ◯ 5 T.

D.

7 tsp. ◯ 1 T.

E.

10 tsp. ◯ 3 T.

Word Problem

F. Rewrite the recipe using only teaspoons.

Fudge Delight

Put 1 pint of ice cream into a 2 quart bowl. Add 1 ounce of hot fudge sauce and 3 cups of ice cream. Add 2 tablespoons of chopped walnuts.

_____

_____

```
1 tablespoon = 3 teaspoons
   1 ounce = 6 teaspoons
      1 cup = 32 teaspoons
     1 pint = 64 teaspoons
    1 quart = 192 teaspoons
```

Create your own math problem and explain your solution.

Math Problem:

_____

Explanation/Solution:

_____

_____

Answers: A. < B. = C. < D. > E. > F. Put 64 teaspoons into a 384-teaspoon bowl. Add 6 teaspoons of hot fudge sauce and 96 teaspoons of ice cream. Add 6 teaspoons of chopped walnuts.

Warm Up

Name: _____

Date: _____

Compare the amounts using the <, >, or = symbols. Remember, 4 quarts equal 1 gallon.

A.

12 quarts ◯ 9 gallons

B.

40 quarts ◯ 3 gallons

C.

16 quarts ◯ 4 gallons

D.

32 quarts ◯ 7 gallons

Word Problem

E. In the chart, convert the quarts to the nearest gallon. Then answer the questions. Remember, 4 quarts equal 1 gallon.

Country	Milk Consumed	Gallons
1. Finland	162 quarts	_____
2. Iceland	160 quarts	_____
3. Ireland	164 quarts	_____
4. Norway	158 quarts	_____
5. Ukraine	141 quarts	_____

6. In which country do the people drink the most milk? _____

7. In which country do the people drink the least amount of milk? _____

8. What is the average amount of milk consumed by the people in those five countries? _____

Create your own math problem and explain your solution.

Math Problem:

Explanation/Solution:

Answers: A. < B. > C. = D. > E1. 41 gallons E2. 40 gallons E3. 41 gallons E4. 40 gallons E5. 35 gallons E6. Ireland E7. Ukraine E8. 157 quarts or 39 gallons

Name: _____

Date: _____

Warm Up

Circle the answer.

A. 20
    × 3
    ___

   50   60

   70   80

B. 50
    × 9
    ___

   40   50

  140   450

C. 90
    × 2
    ___

   70   90

  130   180

D. 40
    × 8
    ___

  120   160

  320   480

Word Problem

E. Eli used an egg carton to store his seed collection. He put 12 seeds in each egg space. How many seeds are in the egg carton?

1. Before you can solve this problem, what important piece of information do you need to know?

_____

2. The egg carton holds 12 eggs. Solve the problem.

_____

Create your own math problem and explain your solution.

Math Problem:

Explanation/Solution:

Answers: A. 60 B. 450 C. 180 D. 320 E1. How many egg spaces are in an egg carton? E2. 12 × 12 = 144 seeds

Name: _____   **Date:** _____

Warm Up

Add the missing factor to make each question true.

A.

9 × 9 ⟨∨⟩ 10 × _____

B.

6 × 3 ⟨∧⟩ 8 × _____

C.

8 × 4 ⟨∨⟩ 3 × _____

D.

8 × 5 ⟨∨⟩ 3 × _____

Word Problem

E. Warwick can count 37 widgets every minute. Sapphire can count 75 widgets every two minutes. Who can count more widgets in ten minutes?

Create your own math problem and explain your solution.

Math Problem:

Explanation/Solution: _____

Answers: A. Any number between 1 and 8 B. Either 1 or 2 C. Any number 11 or greater D. Any number 14 or greater E. Warwick can count 37 x 10 = 370 widgets. Sapphire can count 75 x 5 = 375 widgets. Sapphire can count more widgets than Warwick.

Warm Up

Name: _____

Date: _____

Solve.

A. Four pens cost 20¢. How much does 1 pen cost?

_____

B. Three pieces of gum cost 21¢. How much would 1 piece of gum cost?

_____

C. Five pieces of licorice cost 50¢. How much does 1 piece cost?

_____

D. Two candy bars cost 90¢. How much would 1 candy bar cost?

_____

Word Problem

F. Which one is the better bargain?

Candy Sale #1

4 pieces for 36¢

Candy Sale #2

5 pieces for 40¢

Create your own math problem and explain your solution.

Math Problem: _____

Explanation/Solution: _____

Answers: A. 5¢ B. 7¢ C. 10¢ D. 45¢ E. Candy Sale #1 = 9¢, Candy Sale #2 = 8¢ Candy Sale #2 is the better bargain.

Name:

Date:

Warm Up

Divide. Multiply to check the answer.

A. There are 45 apartments in a 9 story building. How many apartments are on each floor?

_____

B. There are 36 people in 9 apartments. How many people live in each apartment?

_____

C. There are 30 windows in 10 apartments. How many windows are in each apartment?

_____

Word Problem

D. Draw the picture. Write the division problem. Multiply to check the answer.

There are 18 doorbells in 3 rows. How many doorbells are in each row?

Create your own math problem and explain your solution.

Math Problem:

Explanation/Solution:

_____

Warm Up

Write the factors for each number.

A. 21 _____

B. 81 _____

C. 60 _____

D. 17 _____

E. 10 _____

Word Problem

F. Find the least common multiple for the numbers 5 and 7.

5: _____

7: _____

Create your own math problem and explain your solution.

Math Problem:

Explanation/Solution:

_____

Name: _____ **Date:** _____

Warm Up

Estimate by rounding the larger factor. Rewrite the problem and solve.

A. 2 x 94 _____

B. 51 x 6 _____

C. 76 x 5 _____

D. 4 x 78 _____

E. 3 x 83 _____

Word Problem

F. Estimate by rounding the larger factor. Rewrite the problem and solve.

Arnold lifts weights for exercise. Ten times in a row he lifted 92 pounds. How many pounds did Arnold lift in all?

Create your own math problem and explain your solution.

Math Problem: _____

Explanation/Solution:

Answers: A. 2 x 90 = 180 B. 50 x 6 = 300 C. 80 x 5 = 400 D. 4 x 80 = 320 E. 3 x 80 = 240 F. 10 x 90 = 900 pounds

Name: _____

Date: _____

Warm Up

Multiply.

A.
$$\begin{array}{r} \$4.43 \\ \times\ 6 \\ \hline \end{array}$$

B.
$$\begin{array}{r} \$6.21 \\ \times\ 4 \\ \hline \end{array}$$

C.
$$\begin{array}{r} \$7.93 \\ \times\ 10 \\ \hline \end{array}$$

D.
$$\begin{array}{r} \$1.05 \\ \times\ 3 \\ \hline \end{array}$$

E.
$$\begin{array}{r} \$8.75 \\ \times\ 5 \\ \hline \end{array}$$

Word Problem

F. How much would Morgan earn if she cleaned 6 small windows and 5 large windows?

Window Cleaning Special	
Small Windows	$0.74 each
Large Windows	$2.78 each

Create your own math problem and explain your solution.

Math Problem:

Explanation/Solution:

_____

116

Name: _____ **Date:** _____

Warm Up

Find the lowest common denominator for each set of fractions.

A.

$\dfrac{1}{3}$ $\dfrac{1}{10}$

The lowest common denominator is ___.

B.

$\dfrac{4}{12}$ $\dfrac{7}{9}$

The lowest common denominator is ___.

C.

$\dfrac{2}{6}$ $\dfrac{11}{18}$

The lowest common denominator is ___.

D.

$\dfrac{7}{15}$ $\dfrac{4}{5}$

The lowest common denominator is ___.

E.

$\dfrac{6}{14}$ $\dfrac{1}{2}$

The lowest common denominator is ___.

Word Problem

F. Read the word problem and write the math sentence. Find the answer.

Kelly caught 36 fish during the fishing contest. If 1/6 of the fish were green, 2/3 of the fish were gray, and the rest were silver, how many of each color fish did Kelly catch?

Create your own math problem and explain your solution.

Math Problem: _____

Explanation/Solution: _____

Name: _____   **Date:** _____

Warm Up

Find the lowest common denominator for each set of fractions.

A.

$\frac{2}{4}$ $\frac{1}{5}$

The lowest common denominator is ____.

B.

$\frac{1}{6}$ $\frac{2}{3}$

The lowest common denominator is ____.

C.

$\frac{3}{8}$ $\frac{1}{2}$

The lowest common denominator is ____.

D.

$\frac{6}{7}$ $\frac{1}{3}$

The lowest common denominator is ____.

E.

$\frac{5}{6}$ $\frac{4}{8}$

The lowest common denominator is ____.

Word Problem

F. Read the word problem and write the math sentence. Find the answer.

Kiel had 20 acres of farmland. He grew corn on 1/5 of the acres, pumpkins on 1/4 of the acres, and peas on the remaining acres. How many acres of peas did Kiel grow?

Create your own math problem and explain your solution.

Math Problem: _____

Explanation/Solution: _____

Answers: A. 20 B. 6 C. 8 D. 21 E. 24 F. 4 acres of corn, 5 acres of pumpkins, 11 acres of peas

Warm Up

Divide each shape. Shade to show the fraction. Circle the larger fraction.

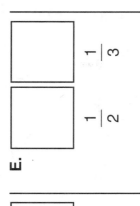

A.
$\frac{1}{9}$ $\frac{8}{9}$

B.
$\frac{1}{10}$ $\frac{6}{8}$

C.
$\frac{2}{5}$ $\frac{3}{7}$

D.
$\frac{6}{10}$ $\frac{3}{4}$

E.
$\frac{1}{2}$ $\frac{1}{3}$

Word Problem

F. Follow the directions.

Dad brought home a super large pizza.

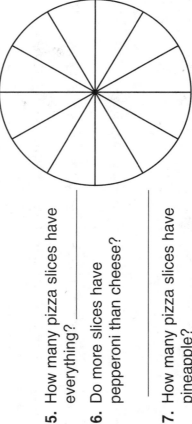

1. 1/4 of the pizza had pepperoni. Label it on the graph.

2. 1/3 of the pizza had just cheese. Label it on the graph.

3. 1/6 of the pizza had pineapple. Label it on the graph.

4. The rest of the pizza had everything. Label it on the graph.

5. How many pizza slices have everything? _____

6. Do more slices have pepperoni than cheese? _____

7. How many pizza slices have pineapple? _____

Create your own math problem and explain your solution.

Math Problem:

Explanation/Solution:

_____

Answers: A.–E. Students need to divide each square into the correct number of parts and color the appropriate sections. A. 8/9 B. 6/8 C. 3/7 D. 3/4 E. 1/2 F1. Label pepperoni to 3 slices. F2. Label cheese to 4 slices. F3. Label pineapple to 2 slices. F4. Label everything to 3 pieces. F5. 3 slices F6. No, more cheese F7. 2 slices

Name: _____   **Date:** _____

Warm Up

Shade. Circle the larger fraction.

A.

$\dfrac{7}{8}$ $\dfrac{3}{10}$

B.

$\dfrac{3}{8}$ $\dfrac{5}{7}$

C.

$\dfrac{4}{6}$ $\dfrac{2}{5}$

D.

$\dfrac{1}{3}$ $\dfrac{5}{9}$

E.

$\dfrac{6}{8}$ $\dfrac{1}{10}$

Word Problem

F. Alistar and Sandra had a contest. Alistar sold 3/10 of his tickets. Sandra sold 1/5 of her tickets. Who sold more tickets?

Create your own math problem and explain your solution.

Math Problem:

Explanation/Solution:

Answers: A. 7/8 B. 5/7 C. 4/6 D. 5/9 E. 6/8 F. Alistar sold more tickets.

118

Warm Up

Name: _____ **Date:** _____

Rewrite each fraction. Add.

A.

$$\frac{3}{5} + \frac{1}{15} \longrightarrow \underline{} + \underline{} = \underline{}$$

B.

$$\frac{1}{8} + \frac{2}{24} \longrightarrow \underline{} + \underline{} = \underline{}$$

Word Problem

C. Oliver had 1/4 of a Soft and Chewy bar, 1/3 of a Smooth and Creamy candy bar, and 1/6 of a Peanut Butter Crunch bar. Does Oliver have more than one candy bar?

Create your own math problem and explain your solution.

Math Problem:

Explanation/Solution: _____

Answers: A. 9/15 + 1/15 = 10/15 B. 3/24 + 2/24 = 5/24 C. No. Oliver had 9/12 or 3/4 of a candy bar.

119

Name: _____ **Date:** _____

Warm Up

Rewrite each fraction. Subtract.

A.

$$\frac{5}{8} - \frac{2}{4} = \frac{}{} - \frac{}{} \longrightarrow \frac{}{} = \frac{}{}$$

B.

$$\frac{8}{9} - \frac{3}{5} = \frac{}{} - \frac{}{} \longrightarrow \frac{}{} = \frac{}{}$$

Word Problem

C. The tub was filled until it was 1/2 full. Joanie drained off 2/6 of the water. How much water was left in the tub?

Create your own math problem and explain your solution.

Math Problem:

Explanation/Solution:

_____

126.

Warm Up

Name: _____

Date: _____

Rewrite each fraction. Subtract.

A.

$$\frac{3}{4} - \frac{1}{8} = \underline{\,-\,-\,-\,}$$

B.

$$\frac{5}{6} - \frac{1}{12} = \underline{\,-\,-\,-\,}$$

Word Problem

C. There was 5/8 of a bag of potting soil. Darrel used 1/4 of the potting soil in the flower pot. How much soil is left?

Create your own math problem and explain your solution.

Math Problem:

Explanation/Solution: _____

121

Warm Up

Rewrite each fraction. Add.

A.

$$\dfrac{1}{2} + \dfrac{1}{4} = \dfrac{}{} + \dfrac{}{} = \dfrac{}{}$$

B.

$$\dfrac{1}{3} + \dfrac{2}{6} = \dfrac{}{} + \dfrac{}{} = \dfrac{}{}$$

Word Problem

C. Denise had 1/2 of a dollar. Louisa had 1/4 of a dollar. Who had more money?

Create your own math problem and explain your solution.

Math Problem:

Explanation/Solution:

_____

Name: _____

Date: _____

Warm Up

Write the number of feet. (Remember, 12 inches = 1 foot.)

A. 12 inches

_____ foot

B. 48 inches

_____ feet

C. 36 inches

_____ feet

D. 84 inches

_____ feet

E. 96 inches

_____ feet

Word Problem

F. Complete the table.

Inches (")	6"	12"		24"		42"	
Feet (')	1/2'		1 1/2'				5 1/2'

1. How many inches are in 4 1/2'? _____

2. How many feet are in 30"? _____

3. How many feet are in 6"? _____

4. How many inches are in 6'? _____

Create your own math problem and explain your solution.

Math Problem:

Explanation/Solution:

_____

Warm Up

To convert pounds (lb.) to ounces (oz.), multiply each weight by 16.

A. The albatross weighs 34 pounds or _____ ounces.

B. The great bustard weighs 40 pounds or _____ ounces.

C. The mute swan weighs 36 pounds or _____ ounces.

D. The trumpeter swan weighs 37 pounds or _____ ounces.

E. The whooper swan weighs 34 pounds or _____ ounces.

Word Problem

F. In the chart, convert each weight in pounds to ounces or from ounces to pounds. Then answer the questions.

Person	Pounds (lb.)	Ounces (oz.)
1. Amir	50	
2. Knox		960
3. Orenda	30	
4. Eva		640

5. Who lifted more weight than Amir? _____

6. Who lifted less weight than Eva? _____

7. If Knox were able to lift 25% more weight, how much weight would he be able to lift? _____

Create your own math problem and explain your solution.

Math Problem:

_____

Explanation/Solution:

Answers: A. 544 oz. B. 640 oz. C. 576 oz. D. 592 oz. E. 544 oz. F1. 800 oz. F2. 60 lb. F3. 480 oz. F4. 40 lb. F5. Knox F6. Orenda F7. 60 lb. x 1/4 or (25%) = 15 lb. 15 + 60 = 75 lb. or 960 oz. x 1/4 (25%) = 240 oz. 240 + 960 = 1200 oz.

Name: _____

Date: _____

Warm Up

Convert the number of feet to yards. Divide the number of feet by 3.

A.

72 feet

_____

B.

15 feet

_____

C.

45 feet

_____

D.

96 feet

_____

E.

18 feet

_____

Word Problem

F. In the chart, convert each height from feet to yards. Round each yard to the nearest whole number. Then answer the questions.

Building	Height (ft.)	Height (yd.)
1. Empire State Building	1,250 ft.	_____
2. Jin Mao Building	1,255 ft.	_____
3. Petronas Towers	1,483 ft.	_____
4. Sears Tower	1,454 ft.	_____

6. Which building is the tallest? _____

7. Which building is taller than the Jin Mao Building and shorter than the Petronas Towers? _____

8. What is the difference in yards between the Petronas Towers and the Empire State Building? _____

Create your own math problem and explain your solution.

Math Problem:

Explanation/Solution:

Name: _____

Date: _____

Warm Up

Compare each measurement using the >, <, or = symbols. Remember, 3 feet equals 1 yard.

A.

891 ft. ◯ 85 yd.

B.

162 ft. ◯ 44 yd.

C.

8,211 ft. ◯ 4,798 yd.

D.

6,942 ft. ◯ 6,110 yd.

Word Problem

E. In the chart, convert the feet to the nearest yard. Remember, 3 feet equals 1 yard.

Mountains		
Mountain	**Height (ft.)**	**Height (yd.)**
1. K2	28,250	_____
2. Kangchenjunga	28,208	_____
3. Lhotse	27,890	_____
4. Makalu	27,790	_____
5. Mt. Everest	29,035	_____

6. Which mountain is the tallest? _____

7. Which mountain is taller than Makalu and shorter than Kangchenjunga? _____

8. What is the average height of the mountains rounded to the nearest yard? _____

Create your own math problem and explain your solution.

Math Problem: _____

Explanation/Solution: _____

Answers: A. > B. > C. < D. < E1. 9,417 yd. E2. 9,403 yd. E3. 9,297 yd. E4. 9,263 yd. E5. 9,678 yd. E6. Mt. Everest E7. Lhotse E8. 9,412 yd.

Name: _____

Date: _____

Warm Up

Compare the two players using the information from the chart.

A. Cal Ripken
Billy Williams

_____ played more
games than _____.

B. Everett Scott
Lou Gehrig

_____ played more
games than _____.

C. Steve Garvey
Lou Gehrig

_____ played more
games than _____.

D. Cal Ripken
Everett Scott

_____ played fewer
games than _____.

E. Billy Williams
Everett Scott

_____ played fewer
games than _____.

Player	Consecutive Games Played
Cal Ripken	2,632
Lou Gehrig	2,130
Everett Scott	1,307
Steve Garvey	1,207
Billy Williams	1,117

Word Problem

F.

1. Which two players combined played in almost as many consecutive games as Cal Ripken? _____

2. What was the average number of consecutive games played by the players? _____

(Information from *Scholastic Book of Records.* 2001. Page 221.)

Create your own math problem and explain your solution.

Math Problem: _____

Explanation/Solution:

Name: _____   **Date:** _____

Warm Up

Compare the amounts using <, >, or = symbols. Remember, 1 ton equals 2,000 pounds.

A.

3,899 lb. ◯ 5 tons

_____

B.

5,669 lb. ◯ 8 tons

_____

C.

4,218 lb. ◯ 2 tons

_____

D.

6,000 lb. ◯ 3 tons

_____

Word Problem

E. In the chart, convert the weight in tons to pounds. Remember, 1 ton equals 2,000 pounds. Then answer the questions below.

Rice Production		
Country	**Amount**	**Pounds**
1. Bangladesh	31 tons	_____
2. China	212 tons	_____
3. India	135 tons	_____
4. Indonesia	51 tons	_____
5. Vietnam	30 tons	_____

6. Which country produces the most rice? _____

7. What is the average amount of rice produced in pounds? _____

8. Which three countries combined produce about the same amount of rice as China? _____
_____

Create your own math problem and explain your solution.

Math Problem:

Explanation/Solution:

_____

Name: _____   **Date:** _____

Warm Up

Find the number for each set of items.

A. 1/3 of 15 scarves _____

B. 1/2 of 20 mittens _____

C. 1/4 of 8 ski caps _____

D. 1/5 of 20 boots _____

E. 1/8 of 2 dozen snow shoes _____

Word Problem

F. There was a great sale at Ski World. All skis were 1/4 off the listed price. What is the sale price for a pair of $40.00 skis? How much money would be saved?

Create your own math problem and explain your solution.

Math Problem:

Explanation/Solution:

_____

126

Name: _____   **Date:** _____

Warm Up

Write the length of time in minutes.

A. 1/2 of an hour

_____

B. 1/4 of an hour

_____

C. 1/3 of an hour

_____ _____

D. 1/10 of an hour

_____

E. 3/4 of an hour

_____

Word Problem

F. Elroy was cooking dinner. The recipe said, "Cook for 2/3 of an hour, turn the food, and cook for another 1/4 of an hour . . . ." How long should Elroy cook the dinner?

Create your own math problem and explain your solution.

Math Problem:

Explanation/Solution:

_____

Name: _____ Date: _____

Warm Up

Find the number of items.

A. 2/3 of 60 raisins

_____

B. 3/5 of 30 trees

_____

C. 3/4 of 80 tractors

_____

D. 5/6 of 90 orchards

_____

E. 3/10 of 100 acres

_____

Word Problem

F. At the Garden of Avon, Farmer Avon has 75 fruit trees. This year 2/5 of the fruit trees produced rotten fruit. How many of trees produced rotten fruit?

Create your own math problem and explain your solution.

Math Problem:

Explanation/Solution:

_____

Answers: A. 40 raisins B. 18 trees C. 60 tractors D. 75 orchards E. 30 acres F. 30 trees

Name: _____ **Date:** _____

Warm Up

Rewrite each improper fraction as a mixed number or whole number.

A.
8/3 = _____

B.
7/2 = _____

C.
3/2 = _____

D.
12/4 = _____

E.
2/1 = _____

F.
14/6 = _____

G.
18/10 = _____

H.
12/5 = _____

I.
23/8 = _____

J.
11/9 = _____

Word Problem

K. Write the problem and solve. Rewrite the improper fraction as a mixed fraction.

Silva ate 1/3 of a hamburger, 2/3 of a cookie, 1/3 of a salad, and 1/3 of a dinner roll. How much food did Silva eat in all?

Create your own math problem and explain your solution.

Math Problem: _____

Explanation/Solution: _____

Answers: *A. 2 2/3 B. 3 1/2 C. 1 1/2 D. 3 E. 2 F. 2 2/6 or 2 1/3 G. 1 8/10 or 1 4/5 H. 2 2/5 I. 2 7/8 J. 1 2/9 K. 1/3 + 2/3 + 1/3 + 1/3 = 5/3 = 1 2/3 of the food*

Name: _____

Date: _____

Warm Up

Subtract each set of mixed numbers. Simplify the answer.

A. 10 3/6 − 3 4/12 = _____

B. 5 2/3 − 2 4/9 = _____

C. 1 2/8 − 1/4 = _____

D. 1 1/3 − 1 1/6 = _____

E. 5 1/2 − 1 4/9 = _____

Word Problem

F. Moses had 3/4 of a dozen farm fresh eggs left to sell. Mr. Hornby visited Moses and bought 1/4 of those eggs. How many eggs does Moses have left? How many eggs did Mr. Hornby buy?

Create your own math problem and explain your solution.

Math Problem:

Explanation/Solution: _____

Answers: A. 7 1/6 B. 3 2/9 C. 1 D. 1/6 E. 4 1/18 F. Moses has 1/2 of a dozen or 6 eggs. Mr. Hornby bought 3 eggs.

132

Name: _____

Date: _____

Warm Up

Add each set of mixed numbers. Simplify the answer.

A. 2 6/9 + 6 8/9 =

_____

B. 1 4/7 + 8 5/7 =

_____

C. 6 2/4 + 10 1/4 =

_____

D. 8 5/8 + 4 7/8 =

_____

E. 2 3/10 + 1 4/10 =

_____

Word Problem

F. Isadora loves to garden. She went to the local nursery and bought 2 1/2 bags of potting soil, 3 1/2 bags of mulch, 6 1/2 bags of fertilizer, and 4 1/2 bags of small pebbles. How many bags of items did Isadora buy in all?

Create your own math problem and explain your solution.

Math Problem:

Explanation/Solution: _____

134

Name: _____ **Date:** _____

Warm Up

Multiply. Simplify the answer.

A. $2/6 \times 8/10 =$ ___

B. $3/4 \times 7/8 =$ ___

C. $3/5 \times 3/9 =$ ___

D. $3/7 \times 1/2 =$ ___

E. $2/9 \times 3/6 =$ ___

Word Problem

F. Write the equation for each word problem. Solve.

□ = 100 pennies

1st	🪙	🪙	🪙	🪙	🪙	🪙
2nd						🪙
3rd			🪙	🪙	🪙	🪙
4th				🪙	🪙	🪙
5th					🪙	🪙

1. Half as many pennies as the 1st graders ___

2. One-fourth the number of pennies as the 5th graders ___

3. Four-tenths the number of pennies as the 3rd graders ___

4. Two-fifths the number of pennies as the 2nd graders ___

5. One-sixth the number of pennies as the 4th graders ___

6. What is half of the total number of pennies collected? ___

Create your own math problem and explain your solution.

Math Problem:

_____

Explanation/Solution:

_____

Answers: A. 4/15 B. 21/32 C. 1/5 D. 3/14 E. 1/9 F1. $1/2 \times 600 = 300$ F2. $1/4 \times 200 = 50$ F3. $4/10 \times 500$ or $2/5 \times 500 = 200$ F4. $2/5 \times 100 = 40$ F5. $1/6 \times 300 = 50$ F6. $1,700 \times 1/2 = 850$

Name: _____

Date: _____

Warm Up

Use the >, <, or = symbols to compare each set of fractions.

A.

1/3 of 15 ◯ 1/4 of 16

B.

4/5 of 25 ◯ 2/5 of 40

C.

1/9 of 9 ◯ 2/3 of 3

D.

5/7 of 35 ◯ 4/5 of 35

E.

3/8 of 24 ◯ 2/3 of 24

Word Problem

F. Jerome's strawberry plants made enough strawberries to fill 132 baskets. Frances' strawberry plants made only enough strawberries to fill 1/4 of the baskets that Jerome's plants did. How many baskets of strawberries did Frances' strawberry plants produce?

Create your own math problem and explain your solution.

Math Problem:

Explanation/Solution: _____

Answers: A. > B. > C. < D. < E. < F. 33 baskets of strawberries

Name: _____ **Date:** _____

Warm Up

Multiply. Simplify the answer.

A.

$5 \times 4/8 =$

_____

B.

$4 \times 3/7 =$

_____

C.

$7 \times 1/4 =$

_____

D.

$3 \times 3/4 =$

_____

E.

$7 \times 7/9 =$

_____

Word Problem

F. The pancake recipe serves 8 and requires 4 cups of flour. If Magda wants to make 1/4 the number of servings, how many cups of flour should she use?

Answers: A. 2 1/2 B. 1 5/7 C. 1 3/4 D. 2 1/4 E. 5 4/9 F. 4 × 1/4 = 4/4 or 1 cup of flour

Create your own math problem and explain your solution.

Math Problem:

Explanation/Solution:

Name: _____

Date: _____

Warm Up

Divide. Simplify the answer.

A.

$6/8 \div 3/10 =$

_____

B.

$4/5 \div 2/10 =$

_____

C.

$3/9 \div 1/5 =$

_____

D.

$2/3 \div 3/5 =$

_____

E.

$9/10 \div 2/9 =$

_____

Word Problem

F. Write the problem. Solve.

Ramona exercises 3 times a day for a total of 1/2 an hour each day. How much time does Ramona exercise during each session?

_____

Create your own math problem and explain your solution.

Math Problem:

Explanation/Solution:

Name: _____

Date: _____

Warm Up

Circle the number.

A. nine thousand, seven hundred nine	**B.** eighty-one thousand, five hundred thirty-one	**C.** forty thousand, four hundred ninety-one	**D.** ten thousand, three hundred eighty-six
97,009	81,000,500,31	40,491	10,386
900,709	81,500,31	404,0091	10,000,300,86
9,709	81,531	4,091	103,00086

Word Problem

E. The population of Smallville has doubled over the last five years. The current population is sixty-seven thousand, one hundred twelve. What was the approximate population five years ago? (Circle the answer.)

three hundred forty thousand   thirty-four thousand

three hundred four thousand   thirty thousand, four hundred

Create your own math problem and explain your solution.

Math Problem:

Explanation/Solution: _____

Name: _____

Date: _____

Warm Up

Round each number.

A. Round to the nearest ten.	B. Round to the nearest hundred.	C. Round to the nearest thousand.	D. Round to the nearest ten thousand.	E. Round to the nearest hundred thousand.
692	158	2,276	76,317	459,878
_____	_____	_____	_____	_____

Word Problem

Presidential Votes

George H.W. Bush	48,886,097
Bill Clinton	44,909,889
Richard Nixon	47,169,911
Ronald Reagan (1984)	54,455,075
Ronald Reagan (1980)	43,899,248

F.

1. Which person had the greatest number of popular votes twice? _____

2. Which years did that person win the presidency? _____

3. Which person had fewer votes than Bill Clinton? _____

4. Which person had more votes than George H.W. Bush? _____

5. What was the difference in votes for Ronald Reagan's two presidential wins? _____

Create your own math problem and explain your solution.

Math Problem: _____

Explanation/Solution: _____

Answers: A. 690 B. 200 C. 2,000 D. 80,000 E. 500,000 F1. Ronald Reagan F2. 1980, 1984 F3. Ronald Reagan F4. Ronald Reagan F5. 10,555,827

14

Name: _____ **Date:** _____

Warm Up

Round each number to the nearest million.

A.

3,258,149

_____

B.

9,442,541

_____

C.

1,647,161

_____

D.

6,948,568

_____

Word Problem

E. Seventeen million, three hundred sixty-eight thousand people visited the Tokyo Disneyland in Japan. Eleven million, seven hundred ninety-six thousand, seven hundred fifty people visited the EPCOT Center in Disney World. How many more people visited Tokyo Disneyland than the EPCOT Center?

Create your own math problem and explain your solution.

Math Problem:

Explanation/Solution:

141

Name: _____   Date: _____

Warm Up

Write the place value of the underlined digit.

A.

9,887_2_77

_____

B.

5,310,11_1_

_____

C.

_1_,628,628

_____

D.

3,9_9_1,562

_____

E.

6,5_8_6,178

_____

Word Problem

F. In 1998, 32,368,301 people lived in California and 7,486,242 people lived in Georgia. How many more people lived in California than in Georgia?

Create your own math problem and explain your solution.

Math Problem:

Explanation/Solution:

Name: _____ **Date:** _____

Warm Up

Rewrite each number in standard form.

A. One hundred nineteen billion, seven hundred million dollars is spent on importing cars.

_____

B. Seventy-nine billion, four hundred million dollars is spent on importing machinery.

_____

C. Thirty-seven billion, three hundred million dollars is spent on importing crude oil.

_____

D. Fifty-three billion, seven hundred million dollars is spent on importing clothes.

_____

Word Problem

E. Find the value of each denomination that is in circulation. Then answer the questions.

Bill	In Circulation	Value	Bill	In Circulation	Value
$1	6,723,236,225	_____	$100	3,527,894,018	_____
$2	548,608,768	_____	$500	287,515	_____
$5	1,546,663,467	_____	$1,000	167,033	_____
$10	1,353,493,806	_____	$5,000	351	_____
$20	4,351,869,983	_____	$10,000	344	_____
$50	998,342,803	_____			

1. Which denomination has the most value of money in circulation?

_____

2. Which denomination has the least value of money in circulation?

_____

Create your own math problem and explain your solution.

Math Problem:

Explanation/Solution:

143

Name: _____ Date: _____

Warm Up

Rewrite each amount of money in standard form.

A. $15/100

_____

B. $44/100

_____

C. $113/100

_____

D. $282/100

_____

E. $3/100

_____

Word Problem

F. Vicky has three piggy banks filled with pennies. In the first piggy bank, she has 62 pennies. In the second piggy bank, she has 95 pennies. In the third piggy bank, she has 214 pennies. How many pennies does Vicky have in all? How much money is that? Write the answers in standard form and as a fraction of a dollar.

Create your own math problem and explain your solution.

Math Problem:

Explanation/Solution:

_____

Name: _____ Date: _____

Warm Up

Write the amounts in order, smallest to greatest.

A. 564,741,221,781

396,243,101,108

810,445,341,081

599,984,798,179

221,221,767,631

_____

_____

_____

_____

_____

Word Problem

B. Find the distance from the sun for each planet.

Mercury is 36 million miles away from the sun.

1. Mars is approximately 4 times farther away from the sun than Mercury. _____
2. Earth is one half the distance of Mars plus 20 million miles. _____
3. Saturn is nine times farther away than Earth plus 59 million miles. _____
4. Uranus is two times farther away than Saturn. _____
5. Pluto is approximately two times farther away than Uranus. _____

Create your own math problem and explain your solution.

Math Problem:

_____

Explanation/Solution:

_____

Answers: A. 221,221,767,631; 396,243,101,108; 564,741,221,781; 599,984,798,179; 810,445,341,081 B1. 144 million miles B2. 92 million miles B3. 887 million miles B4. 1.77 billion miles B5. 3.54 billion miles

Name: _____

Date: _____

Warm Up

Rewrite each number adding the missing commas.

A.

32167435928 8

_____

B.

399227521310

_____

C.

71563108897 6

_____

D.

86553410329 1

_____

Word Problem

E. In the chart, rewrite each population in standard form. Then answer the questions below.

The Population for Each Continent	
1. Africa	778 million _____
2. Antarctica	0 people _____
3. Asia	3 billion 641 million _____
4. Australia	30 million _____
5. Europe	727 million _____
6. N. America	476 million _____
7. S. America	343 million _____

8. Which continent has the largest population? _____

9. Which continent has the smallest population? _____

10. Which continent has population larger than Antarctica and smaller than South America? _____

11. Which continent has a population smaller than Asia and larger than Europe? _____

12. What is the population of your continent? _____

Create your own math problem and explain your solution.

Math Problem:

Explanation/Solution:

145

Name: _____

Date: _____

Warm Up

Write the amount of money.

A. 1/5 of $1.00

_____

B. 3/4 of $1.00

_____

C. 1/2 of $1.00

_____

D. 2/5 of $1.00

_____

E. 1/4 of $1.00

_____

Word Problem

F. Rewrite the amounts of money as fractions of a dollar. Solve the problem.

Gracie had 3 dimes, 6 pennies, 2 quarters and 1 nickel. How much money does Gracie have?

Create your own math problem and explain your solution.

Math Problem:

Explanation/Solution: _____

_____

Answers: A. $0.20 or 20¢ B. $0.75 or 75¢ C. $0.50 or 50¢ D. $0.40 or 40¢ E. $0.25 or 25¢ F. 30/100 + 6/100 + 50/100 + 5/100 = 91/100, 91¢ or $0.91

147

Warm Up

Find the number.

A.

10% of 100 socks

_____

B.

15% of 100 ties

_____

C.

45% of 100 shoes

_____

D.

25% of 100 hats

_____

E.

95% of 100 belts

_____

Word Problem

Store #1
Raincoats on Sale!
30% off
Regular Price
$100.00

Store #2
Raincoats on Sale!
20% off of the
Regular Price of
$100.00
Plus an Extra 10%

F. Which one is the better deal?

Create your own math problem and explain your solution.

Math Problem:

Explanation/Solution:

Name: _____ **Date:** _____

Warm Up

Find the number.

A.	B.	C.	D.	E.
10% of 50 board games	20% of 30 card games	50% of 20 pinball games	80% of 10 video games	40% of 30 computer games
_____	_____	_____	_____	_____

Word Problem

F. Gregory has a Crazy Cat's Games coupon. What is the total price for 3 games? (Each game costs $20.00.)

```
Crazy Cat's Games Coupon

    1st Game—Save 10%

    2nd Game—Save 25%

    3rd Game—Save 40%
```

Create your own math problem and explain your solution.

Math Problem:

_____

Explanation/Solution:

_____

Answers: A. 5 board games   B. 6 card games   C. 10 pinball games   D. 8 video games   E. 12 computer games   F. 1st game: $18.00, 2nd game: $15.00, 3rd game: $12.00, total cost: $45.00

146

Name: _____ **Date:** _____

Warm Up

Rewrite each decimal as a percent.

A.
.45 = _____ %

C.
.25 = _____ %

E.
.21 = _____ %

G.
.92 = _____ %

I.
.54 = _____ %

B.
.83 = _____ %

D.
.87 = _____ %

F.
.34 = _____ %

H.
.28 = _____ %

J.
.61 = _____ %

Word Problem

K. Find the decimal. Rewrite as a percent.

Sedgwick correctly answer 3 out of the 7 multiple choice questions and 2 out of the 3 true and false questions. What was his total score on the quiz?

Create your own math problem and explain your solution.

Math Problem: _____

Explanation/Solution: _____

Answers: A. 45% B. 83% C. 25% D. 87% E. 21% F. 34% G. 92% H. 28% I. 54% J. 61% K. Sedgwick answered 5 out of 10 questions correctly or .50 or 50%.

Warm Up

150

Rewrite each decimal as a percent.

A. .02 = _____ %	**C.** .065 = _____ %
B. .03 = _____ %	**D.** .033 = _____ %
E. .013 = _____ %	**G.** .017 = _____ %
F. .051 = _____ %	**H.** .079 = _____ %
	I. .081 = _____ %
	J. .049 = _____ %

Word Problem

K. In order to be admitted to the Advanced Learning Committee, Sylvania needs to earn a score of at least 80% on the pre-admission entrance exam. The test has 35 questions. How many questions will Sylvania need to answer correctly?

Create your own math problem and explain your solution.

Math Problem:

_____

Explanation/Solution:

_____

Answers: A. 2% B. 3% C. 6.5% D. 3.3% E. 1.3% F. 5.1% G. 1.7% H. 7.9% I. 8.1% J. 4.9% K. 28 questions

Name: _____ **Date:** _____

Warm Up

Rewrite each decimal as a fraction.

A. .3 _____

B. .2 _____

C. .8 _____

D. .06 _____

E. .01 _____

Word Problem

F. Jasper took a math test. There were 10 problems on the test. Jasper answered 7 of the questions correctly. What was Jasper's score? Write the score as a fraction and as a decimal.

Create your own math problem and explain your solution.

Math Problem:

Explanation/Solution:

Answers: A. 3/10 B. 2/10 C. 8/10 D. 6/100 E. 1/100 F. 7/10, .7

Name: _____ **Date:** _____

Warm Up

152

Write each set of decimals in order—smallest to greatest.

A.

.1 .3 .9 .6

____, ____, ____, ____

B.

.8 .1 .9 .7

____, ____, ____, ____

C.

.4 .1 .2 .8

____, ____, ____, ____

D.

.5 .1 .9 .2

____, ____, ____, ____

E.

.3 .6 .8 .4

____, ____, ____, ____

Word Problem

F. Each slug had one minute to crawl as far as it could. Use the results to answer each question.

Speedy Slugs	
Slugo	.3 centimeters
Pokey	.9 centimeters
Wiggles	.5 centimeters
Bumpy	.2 centimeters
Waldo	.7 centimeters

1. Which slug covered the greatest distance? _____

2. Which slug crawled a shorter distance than Slugo? _____

3. Which slug crawled a longer distance than Slugo but a shorter distance than Waldo? _____

4. What was the average length crawled by the slugs? _____

Create your own math problem and explain your solution.

Math Problem: _____

Explanation/Solution: _____

Name: _____

Date: _____

Warm Up

Rewrite each number in decimal form.

A. eight and seven-tenths

_____

B. ten and eight-tenths

_____

C. one and four-tenths

_____

D. eighty-four and three-tenths

_____

E. forty-three and nine-tenths

_____

Word Problem

F. In the chart, rewrite each money amount in standard form. Then answer the questions below.

Pay Scale in Dollars for After-School Jobs	
delivering newspapers	two-tenths per paper _____
walking pets	seven-tenths per dog _____
babysitter	one and eight-tenths _____
tutor	two and three-tenths _____

1. If Delta spends 4 hours babysitting, how much would she earn? _____

2. If Isaac delivers 74 newspapers, how much would he earn? _____

3. If Gail tutors for 3 hours, how much would she earn? _____

4. If Plato walks 9 dogs, how much would he earn? _____

Create your own math problem and explain your solution.

Math Problem:

Explanation/Solution:

Name: _____ **Date:** _____

Warm Up

Use the > or < symbols to compare each set of numbers.

A.

6.1 ◯ 6.8

B.

3.8 ◯ 3.7

C.

2.1 ◯ 2.9

D.

6.8 ◯ 6.7

E.

7.5 ◯ 7.6

Word Problem

F. In the chart, round the price of each pet supply to the nearest tenth. Then answer the question.

Mitchell has a new dog named Mr. Foozle. Mitchell had $10.54 to spend at the pet store. He needs to buy Mr. Foozle one of everything. Does Mitchell have enough money?

Pet Store	
Brushes	$2.18
Bowls	$3.20
Chew Toys	$0.54
Bones	$0.87
Food	$3.71

_____
_____
_____
_____

Create your own math problem and explain your solution.

Math Problem:

Explanation/Solution:

_____

Name: _____

Date: _____

Warm Up

Write each set of decimals in order—smallest to greatest.

A.

.12 .23 .83 .65

_____, _____, _____, _____

B.

.71 .50 .10 .66

_____, _____, _____, _____

C.

.11 .96 .10 .99

_____, _____, _____, _____

D.

.13 .81 .18 .74

_____, _____, _____, _____

Word Problem

E. Each rabbit had 3 chances to jump. Each rabbit's longest jump was recorded on the scoreboard.

Rabbit Athletes	
Jumper	3.74 meters
Flash	3.91 meters
Snowball	3.84 meters
Hoppy	3.85 meters
Bouncer	3.37 meters

1. Which rabbit jumped the farthest distance? _____

2. Which rabbit jumped a farther distance than Snowball but a shorter distance than Flash? _____

3. Which rabbit jumped less than 3.50 meters? _____

4. What was the average length of the jumps? _____

Create your own math problem and explain your solution.

Math Problem:

Explanation/Solution:

_____

Answers: A. .12, .23, .65, .83 B. .10, .50, .66, .71 C. .10, .11, .96, .99 D. .13, .18, .74, .81 E1. Flash E2. Hoppy E3. Bouncer E4. 3.74 meters

Name: _____ **Date:** _____

Warm Up

Rewrite each number in standard form.

A.

three and seventy-
seven hundredths

_____

B.

six and
nine hundredths

_____

C.

twenty-two and
four hundredths

_____

D.

seven hundredths

_____

E.

ten and sixty-three
hundredths

_____

Word Problem

F. In the chart, rewrite each money amount in standard form. Then answer the questions below.

Weekly Allowances in Dollars	
Bryan	two and thirty-one hundredths _____
Gabby	five and twenty-five hundredths _____
Lorenzo	four and twelve hundredths _____
Juan	six and eighty-seven hundredths _____
Quitzel	eight and sixty-four hundredths _____

1. Who receives the smallest allowance? _____

2. Name the children who receive more than $5 for their allowance. _____

3. Name the children who receive an allowance smaller than Juan's. _____

4. Compare your allowance to Bryan's. _____

Create your own math problem and explain your solution.

Math Problem:

Explanation/Solution:

Answers: A. 3.77 B. 6.09 C. 22.04 D. 0.07 E. 10.63 F. Chart: $2.31, $5.25, $4.12, $6.87, $8.64 F1. Bryan F2. Gabby, Juan, and Quitzel F3. Lorenzo, Gabby, and Bryan F4. Answers will vary.

Name: _____

Date: _____

Warm Up

Use the > or < symbols to compare each set of numbers.

A.

10.761 $\bigcirc$ 10.486

B.

7.486 $\bigcirc$ 7.439

C.

4.478 $\bigcirc$ 4.411

D.

3.556 $\bigcirc$ 3.561

E.

1.106 $\bigcirc$ 1.831

Word Problem

F. To find each player's batting average, divide the number of hits by the number of at bats. Then answer the questions below.

	Hits	At Bats	Batting Average
1. Ginger	4	7	_____
2. Snaps	2	7	_____
3. Dusty	3	7	_____
4. Rolly	1	7	_____

5. Which player has the worst batting average? _____

6. As a player gets more hits, what happens to the batting average? _____

7. In order to bat "one thousand," how many hits would a player need to have? _____

Create your own math problem and explain your solution.

Math Problem:

Explanation/Solution:

_____

Answers: A. > B. > C. > D. < E. < F1. .571 F2. .286 F3. .429 F4. .143 F5. Rolly F6. As the player gets more hits, the average increases. F7. A player would need to get a hit at every at bat. (In baseball terms, "one thousand" means a score of 1.000.)

Name: _____ **Date:** _____

Warm Up

Write the number of tenths.

A. 2.23

_____

B. 3.25

_____

C. 10.16

_____

D. 10.57

_____

E. 1.72

_____

Word Problem

F. Read and solve the word problem.

Echo has nine tenths of a dollar. Caroline has 96¢. Who has more money?

Create your own math problem and explain your solution.

Math Problem:

Explanation/Solution:

_____

Name:

Date:

Warm Up

Circle the number in the hundredths column.

A. 44.131

B. 10.376

C. 10.872

D. 44.623

E. 72.343

F. 2.107

G. .726

H. 1.595

I. .939

J. 275.168

Word Problem

K. Right now Sally has had 2 hits out of 11 at bats. Her batting average is .182. What would happen to her average if she had 3 more hits out of 3 more at bats?

Create your own math problem and explain your solution.

Math Problem:

Explanation/Solution:

Name: _____ **Date:** _____

Warm Up

Identify the place value for each underlined number.

A.
95.<u>1</u>10

_____

B.
<u>1</u>0.583

_____

C.
2.<u>8</u>45

_____

D.
104.99<u>3</u>

_____

E.
2.<u>4</u>42

_____

Word Problem

F. Uncle Ebenezer had $1.00 to divide equally among his nephews. If each nephew were given five hundredths of a dollar, how much money would each nephew receive?

Create your own math problem and explain your solution.

Math Problem:

Explanation/Solution:

Name: _____ **Date:** _____

Warm Up

Find the sales tax for each item or set of items. The sales tax rate is 7%.

A.

sandals $12.00

_____

B.

surf board $20.00

_____

C.

scuba suit $48.00

_____

D.

fins $3.00
scuba mask $5.00

_____

E.

beach towel $9.00
boggie board $32.00

_____

Word Problem

F. The tax rate in Surfin' City is 8%. Fill in the chart. Then answer the question below.

5. Hillary has $20.00 to spend. What items can she buy from Surfin' City Surf Shop. How much change will she receive?

Item	Price	+ Tax	Total Price
1. Swim Suit	$16.00		
2. Life Vest	$11.00		
3. Snorkeling Equipment	$10.00		
4. Waterproof Watch	$12.00		

Create your own math problem and explain your solution.

Math Problem: _____

Explanation/Solution:

Name: _____

Date: _____

Warm Up

If the minimum wage were $6.00, what would happen to it if the following factors were applied to it?

A. an increase of 10¢

_____

B. a pay cut of 10¢

_____

C. a pay cut of 10%

_____

D. a pay raise of 10%

_____

Word Problem

E. Which job pays more?

Job #1	Job #2
Earn $5.00 an hour for a 40-hour work week.	Earn $4.00 an hour for the first 30 hours and $6.00 an hour for the second 10 hours.

Create your own math problem and explain your solution.

Math Problem:

Explanation/Solution: _____

Answers: A. $6.10 B. $5.90 C. $5.40 D. $6.60 E. Job #1 pays more—Job #1 pays $200 a week, Job #2 pays $180 a week.

Name: _____

Date: _____

Warm Up

Use the chart to answer the questions.

A. Did *My Mother Is a House* products make the most amount of money?

_____

B. Which movie products made less than one hundred dollars?

_____

C. Which movie products made more than *Look Out Below!* and less than *It's Hot in the Desert?*

_____

D. What is the total amount of money earned on movie-related products?

_____

Money Made on Movie-Related Products	
My Mother Is a House	$3,781,789,286
Look Out Below!	$643,103
It's Hot in the Desert	$25,967,832,410
Dessert for Breakfast	$8,479
Cacti Are Prickly	$461,010,515,372
Blowfish v. Megamonster	$22

Word Problem

E.

1. Which movies made billions of dollars in products? _____

2. Which movies made less than one billion dollars in products? _____

3. Which movie made the most amount of money? _____
4. Which movie made the least amount of money? _____

Create your own math problem and explain your solution.

Math Problem: _____

Explanation/Solution: _____

Name: _____   Date: _____

Warm Up

Find the tax amount for each item. The sales tax rate is 5%.

A.
boom box
$112
_____

B.
CD player
$49.00
_____

C.
CD collection
$278.00
_____

D.
speakers
$163.00
_____

E.
headphones
$35.00
_____

Word Problem

F. Fill in the chart and answer the question.

4. How much money will Dominic need to buy 2 pairs of shorts, 5 shirts, and 3 pairs of pants?

Better-Buy Clothes

Item	Price	Discount	Sales Price	+ 5% Tax	Total Price
1. shirts	$15.00	20%			
2. pants	$25.00	15%			
3. shorts	$10.00	25%			

Create your own math problem and explain your solution.

Math Problem:

Explanation/Solution:

Answers: A. $5.60 B. $2.45 C. $13.90 D. $8.15 E. $1.75 F1. $12.00 + $.60 = $12.60 F2. $21.25 + $1.06 = $22.31 F3. $7.50 + $0.38 = $7.88 F4. ($7.88 x 2) + ($12.60 x 5) + ($22.31 x 3) = $145.69

Name: _____

Date: _____

Warm Up

Add or subtract. Write the answer as a mixed number.

A.

$3/2 + 4/2 =$

_____

B.

$9/3 + 5/3 =$

_____

C.

$8/5 - 1/5 =$

_____

D.

$6/3 - 1/3 =$

_____

E.

$5/2 + 6/2 =$

_____

Word Problem

F. Write the problem. Write the answer as a mixed number.

Agatha bought 3/2 pounds of sliced ham, 5/2 pounds of salami, and 7/2 pounds of sliced turkey. How many pounds of meat did Agatha buy?

Create your own math problem and explain your solution.

Math Problem:

Explanation/Solution: _____

166

Name: _____

Date: _____

Warm Up

Find the product.

A.

$(7 \times 4) \times 1 =$ _____

B.

$7 \times (4 \times 1) =$ _____

C.

$(9 \times 7) \times 9 =$ _____

D.

$9 \times (7 \times 9) =$ _____

E.

$9 \times (3 \times 1) =$ _____

F.

$(9 \times 3) \times 1 =$ _____

G.

$2 \times (8 \times 4) =$ _____

H.

$(2 \times 8) \times 4 =$ _____

I.

$5 \times (6 \times 0) =$ _____

J.

$(5 \times 6) \times 0 =$ _____

Word Problem

K. Greta bought two bags of lassos. In each bag there are three lassos. Greta made 12 knots in each lasso. How many knots did Greta make in all?

Create your own math problem and explain your solution.

Math Problem: _____

Explanation/Solution: _____

Answers: A. 28 B. 28 C. 567 D. 567 E. 27 F. 27 G. 64 H. 64 I. 0 J. 0 K. $2 \times 3 \times 12 = 72$ knots

167

Warm Up

Solve.

A.

$$\begin{array}{r} 72 \\ \times\ 6 \\ \hline \end{array}$$

B.

$$\begin{array}{r} 66 \\ \times\ 9 \\ \hline \end{array}$$

C.

$$\begin{array}{r} 42 \\ \times\ 8 \\ \hline \end{array}$$

D.

$$\begin{array}{r} 40 \\ \times\ 3 \\ \hline \end{array}$$

E.

$$\begin{array}{r} 51 \\ \times\ 7 \\ \hline \end{array}$$

Word Problem

F. The Flash basketball team has made 87 baskets per game for the last 6 games. How many baskets have they made in all?

_____

Create your own math problem and explain your solution.

Math Problem:

Explanation/Solution: _____

Answers: A. 432 B. 594 C. 336 D. 120 E. 357 F. 87 x 6 = 522 baskets

Name: _____

Date: _____

Warm Up

168

Multiply.

A.

$$\begin{array}{r} 9,187 \\ \times\ 3 \\ \hline \end{array}$$

B.

$$\begin{array}{r} 1,017 \\ \times\ 5 \\ \hline \end{array}$$

C.

$$\begin{array}{r} 3,624 \\ \times\ 9 \\ \hline \end{array}$$

D.

$$\begin{array}{r} 5,621 \\ \times\ 8 \\ \hline \end{array}$$

E.

$$\begin{array}{r} 6,710 \\ \times\ 2 \\ \hline \end{array}$$

Word Problem

F. The fishing boat has five refrigerated compartments. Each compartment can hold 1,048 fish. How many fish in all can be refrigerated?

Create your own math problem and explain your solution.

Math Problem:

Explanation/Solution:

Name: _____ **Date:** _____

Warm Up

Write the missing number.

A.
$9 \times 9 =$ _____
$9 \times 90 =$ _____
$9 \times$ _____ $= 8{,}100$
$9 \times 9{,}000 =$ _____

B.
$2 \times 1 =$ _____
$2 \times 10 =$ _____
$2 \times 100 =$ _____
$2 \times$ _____ $= 2{,}000$

C.
$3 \times 3 =$ _____
$3 \times$ _____ $= 90$
$3 \times 300 =$ _____
$3 \times 3{,}000 =$ _____

D.
$5 \times 5 =$ _____
$5 \times 50 =$ _____
$5 \times 500 =$ _____
$5 \times 5{,}000 =$ _____

Word Problem

E.

1. Jenny filled 16 bags with 50 plastic cups. How many plastic cups did Jenny put into the bags? _____

2. Jamie filled 160 bags with 50 plastic cups. How many plastic cups did Jamie put into the bags? _____

3. Jett filled 16 bags with 500 plastic cups. How many cups did Jett put into the bags? _____

4. Who used the greatest number of cups? _____

5. Who used the fewest number of cups? _____

Create your own math problem and explain your solution.

Math Problem:

Explanation/Solution:

_____

Answers: A. 81; 810; 900; 81,000 B. 2; 20; 200; 1,000 C. 9; 30; 900; 9,000 D. 25; 250; 2,500; 25,000 E1. 800 E2. 8,000 E3. 8,000 E4. Jamie and Jett E5. Jenny

Warm Up

Name: _____ **Date:** _____

Circle the prime number.

A. _____

10   3

B. _____

6   5

C. _____

2   4

D. _____

17   24

E. _____

38   83

Word Problem

F. Using the chart, read the clues to discover the mystery prime number.

The mystery prime number has the following:

- a 3 as one of the digits
- has two odd numbers as digits
- a digit in the ones place that is a larger number than the digit that is in the tens place

What is the mystery number? _____

11	12	13	14	15	16
21	22	23	24	25	26
31	32	33	34	35	36
41	42	43	44	45	46

Create your own math problem and explain your solution.

Math Problem:

Explanation/Solution:

Warm Up

Name: _____ **Date:** _____

Circle the composite number.

A.

10   1

B.

3   9

C.

6   5

D.

9   2

E.

13   15

Word Problem

F. Circle the composite numbers. Write the factors for each number.

1	2	3	4	5	6
11	12	13	14	15	16
21	22	23	24	25	26
31	32	33	34	35	36
41	42	43	44	45	46

Create your own math problem and explain your solution.

Math Problem:

Explanation/Solution:

Name: _____ **Date:** _____

Warm Up

Write the factors for each number. Circle "Prime" or "Composite."

A. 39 _____

Prime   Composite

B. 51 _____

Prime   Composite

C. 47 _____

Prime   Composite

D. 91 _____

Prime   Composite

Word Problem

E. Find the mystery number.

The mystery number has the following characteristics:

- I am an odd number.
- When the two digits are multiplied, the product is larger than 40.
- When one digit is divided by the other digit, the quotient is 1.

What is the mystery number? _____

47	48	49	50	51	52
57	58	59	60	61	62
67	68	69	70	71	72
77	78	79	80	72	82
87	88	89	90	73	92

Create your own math problem and explain your solution.

Math Problem: _____

Explanation/Solution: _____

173

Warm Up

Circle sets of three. Write the remainder.

A.

Remainder: _____

B.

Remainder: _____

C.

Remainder: _____

D.

Remainder: _____

Word Problem

E. Three people can ride in each car of the roller coaster. How many cars are needed for 14 people?

Create your own math problem and explain your solution.

Math Problem:

Explanation/Solution:

174

Name: _____ **Date:** _____

Warm Up

Divide. Write the remainder.

A.

$2\overline{)7}$

Remainder: _____

B.

$3\overline{)5}$

Remainder: _____

C.

$4\overline{)11}$

Remainder: _____

D.

$5\overline{)13}$

Remainder: _____

E.

$3\overline{)10}$

Remainder: _____

Word Problem

F. There are 30 cookies in the bag. Divide the cookies evenly among nine classmates. How many whole cookies will each classmate receive?

Create your own math problem and explain your solution.

Math Problem:

Explanation/Solution: _____

175

Warm Up

Divide. Write the remainder.

A.

9)29

Remainder: _____

B.

6)51

Remainder: _____

C.

4)85

Remainder: _____

D.

3)13

Remainder: _____

E.

7)30

Remainder: _____

Word Problem

F. Roscoe makes delicious miniature cupcakes. He has 44 jellybeans for seven cupcakes. How many jellybeans can Roscoe put on each cupcake?

Create your own math problem and explain your solution.

Math Problem: _____

Explanation/Solution: _____

176

Date: _____

Warm Up

Divide. Multiply and add to check the answer. Write the remainder.

A. There are 17 nets and 9 courts. How many nets are on each court?

_____

B. There are 31 tennis balls and 10 cans. How many tennis balls are in each can?

_____

C. There are 9 tennis players on 2 teams. How many tennis players are on each team?

_____

Word Problem

D. There are 14 tennis rackets and 4 players. How many tennis rackets does each player have?

Create your own math problem and explain your solution.

Math Problem:

Explanation/Solution: _____

Name: _____ **Date:** _____

Warm Up

Divide. Write the remainder.

A.

$7\overline{)37}$

Remainder: _____

B.

$8\overline{)84}$

Remainder: _____

C.

$9\overline{)85}$

Remainder: _____

D.

$7\overline{)46}$

Remainder: _____

E.

$8\overline{)11}$

Remainder: _____

Word Problem

F. Riva has 67 puppy slippers. How many puppies could have a complete set of slippers?

Create your own math problem and explain your solution.

Math Problem:

Explanation/Solution:

_____

Answers: A. 5 R2 B. 10 R4 C. 9 R4 D. 6 R4 E. 1 R3 F. 67 ÷ 4 (paws on a puppy) = 16 (sets) R3

Warm Up

Name: _____ **Date:** _____

Find the average. Round the answer to the nearest whole number.

A. 95, 98, 12 _____

B. 37, 83, 47 _____

C. 151, 826, 299 _____

D. 553, 729, 186 _____

Word Problem

Cookies Sold

	Chocolate	Peanut Butter	Sugar
June	621	953	494
July	536	381	342
August	721	677	510

E. Round the answer to the nearest whole number.

1. What was the total number of cookies sold during the month of June? _____

2. What was the average number of peanut butter cookies sold over three months? _____

3. On average, were there more sugar cookies sold or more chocolate cookies sold? _____

4. What was the average of cookies sold in July? _____

Create your own math problem and explain your solution.

Math Problem:

Explanation/Solution:

_____

Answers: A. 68 B. 56 C. 425 D. 489 E1. 2,068 E2. 670 E3. More chocolate cookies were sold. E4. 420 cookies

Name: _____

Date: _____

Warm Up

Divide. Write the remainder.

A.

$3\overline{)148}$

Remainder: _____

B.

$5\overline{)104}$

Remainder: _____

C.

$7\overline{)855}$

Remainder: _____

D.

$6\overline{)134}$

Remainder: _____

E.

$9\overline{)345}$

Remainder: _____

Word Problem

F. There are 4 jars and 510 paper clips. How many paper clips can be put into each jar to have an equal amount per jar?

_____

Create your own math problem and explain your solution.

Math Problem:

Explanation/Solution:

Answers: A. 49 R1   B. 20 R4   C. 122 R1   D. 22 R2   E. 38 R3   F. 127 R2

Name: _____

Date: _____

Warm Up

Divide. Write the remainder.

A.

$2\overline{)651}$

Remainder: _____

B.

$7\overline{)942}$

Remainder: _____

C.

$9\overline{)962}$

Remainder: _____

D.

$5\overline{)618}$

Remainder: _____

E.

$3\overline{)172}$

Remainder: _____

Word Problem

F. Lucy has 868 stamps in a stamp collection. She needs to buy pages for her new stamp album. Each album page can hold three stamps. How many album pages does Lucy need to buy?

Create your own math problem and explain your solution.

Math Problem:

Explanation/Solution: _____

Warm Up

Name: _____ **Date:** _____

Divide. Write the remainder.

A.

$9\overline{)9{,}429}$

Remainder: _____

B.

$2\overline{)6{,}381}$

Remainder: _____

C.

$6\overline{)7{,}844}$

Remainder: _____

D.

$7\overline{)6{,}914}$

Remainder: _____

E.

$8\overline{)8{,}309}$

Remainder: _____

Word Problem

F. There are 9 commuter trains for 7,534 commuters. Each car can transport 719 commuters. How many commuters can ride the train to work? How many commuters won't be able to get on one of the commuter trains?

Create your own math problem and explain your solution.

Math Problem:

Explanation/Solution: _____

Answers: A. 1,047 R6   B. 3,190 R1   C. 1,307 R2   D. 987 R5   E. 1,038 R5   F. The train can carry 6,471 people. 1,063 people will not be able to ride on the train.

Name: _____

Date: _____

Warm Up

Add parentheses to make each equation true.

A.

31 − 9 + 9 = 31

_____

B.

31 − 9 + 9 = 13

_____

C.

31 − 9 − 9 = 13

_____

D.

31 + 9 − 9 = 31

_____

E.

31 − 9 ÷ 3 = 28

_____

Word Problem

F. Write the equation.

Fiona gathered 108 pebbles, sold 22, and collected 56 more which she also sold and then she collected 3 more pebbles. Fiona now has 33 pebbles.

_____

Create your own math problem and explain your solution.

Math Problem:

Explanation/Solution: _____

Name: _____

Date: _____

Warm Up

Add parentheses to make each equation true.

A.

$64 - 3 + 4 = 65$

B.

$64 - 3 + 4 = 57$

C.

$64 + 3 - 4 = 63$

D.

$64 + 3 + 4 = 71$

E.

$64 + 3 \times 4 = 76$

Word Problem

F. Write the equation. Solve.

Murdock had 64 scallions. He sold 19 of them and caught 6 more and gave away 6. He now has 33 scallions.

Create your own math problem and explain your solution.

Math Problem:

Explanation/Solution: _____

Answers: A. $(64 - 3) + 4 = 65$ B. $64 - (3 + 4) = 57$ C. $(64 + 3) - 4 = 63$ D. $64 + (3 + 4) = 71$ or $(64 + 3) + 4 = 71$ E. $64 + (3 \times 4) = 76$ F. $64 - (19 + 6) + 6 = 45$

184

Name:

Date:

Warm Up

Convert each centimeter to inches by multiplying the number by .39 inches.

A. 2 cm

_____ in.

B. 9 cm

_____ in.

C. 8 cm

_____ in.

D. 12 cm

_____ in.

E. 20 cm

_____ in.

Word Problem

F. Last year, Bradford grew 14 centimeters. His twin sister, Brenda, grew 6 inches. How much did each grow? Who grew the most? Write the answers in centimeters and in inches. (*Note:* To convert inches to centimeters, multiply the number by 2.54 cm.)

Create your own math problem and explain your solution.

Math Problem:

Explanation/Solution:

_____

Answers: A. .78 in. B. 3.51 in. C. 3.12 in. D. 4.68 in. E. 7.8 in. F. Bradford grew 14 cm or 5.46 in. Brenda grew 15.24 cm or 6 in. Brenda grew the most.

185

Warm Up

Name: _____ **Date:** _____

Convert each meter (m) to nearest whole yards (yd.). Multiply the number of meters by 1.094 yards.

A. 36 meters

_____ yd.

B. 10 meters

_____ yd.

C. 29 meters

_____ yd.

D. 96 meters

_____ yd.

E. 105 meters

_____ yd.

Word Problem

F. In the chart, convert each yard to meters by multiplying the number of yards by .9144 meters. Round the answer to the nearest whole meter. Then answer the question below.

6. To get ready for the next swim meet, Melanie has been swimming 500 meters each day. Which race would be the best one for her to swim in? Why?

Swimming Sign-Ups

1.	100 yards	_____ meters
2.	200 yards	_____ meters
3.	500 yards	_____ meters
4.	1,000 yards	_____ meters
5.	1,500 yards	_____ meters

Create your own math problem and explain your solution.

Math Problem: _____

Explanation/Solution: _____

Name: _____ **Date:** _____

Warm Up

Write the number of kilometers (km) to the nearest whole number. Remember, 1 mile equals 1.609 kilometers.

A.

6 mi. = _____ km

B.

8 mi. = _____ km

C.

2 mi. = _____ km

D.

3 mi. = _____ km

E.

5 mi. = _____ km

F.

1 mi. = _____ km

G.

7 mi. = _____ km

H.

4 mi. = _____ km

I.

9 mi. = _____ km

J.

12 mi. = _____ km

Word Problem

K. Carmen's gas tank can hold 5 gallons of gas. Her car's mileage is 16 miles-per-gallon. Does Carmen have enough gas to go to Knottingham Park and Sherwood Castle and return back to the hotel?

Little Robin's Playhouse	10 km
Friar Chicken Farm	18 km
Maid's Motel	23 km
Knottingham Park	27 km
King's Horses	31 km
Sherwood Castle	35 km

Create your own math problem and explain your solution.

Math Problem:

Explanation/Solution:

Answers: A. 10 km B. 13 km C. 3 km D. 5 km E. 8 km F. 2 km G. 11 km H. 6 km I. 14 km J. 19 km K. Yes. She has enough gas to go 129 km (5 x 16 = 80 x 1.609 = 129 km)

187

Name: _____

Date: _____

Warm Up

Convert each temperature in degrees Celsius (C) to degrees Fahrenheit (F). (Formula: °F = °C x 1.8 + 32)

A. 10°C _____

B. 3°C _____

C. 28°C _____

D. 79°C _____

E. 41°C _____

Word Problem

F. Zaire's mom said he could go fishing if the temperature was cooler than 70°F. The outside temperature is 20°C. Can Zaire go fishing?

_____
_____

Create your own math problem and explain your solution.

Math Problem:

_____

Explanation/Solution:

_____

Answers: A. 50°F B. 37.4°F C. 82.4°F D. 174.2°F E. 105.8°F F. 68°F; Yes, Zaire can go fishing.

188

Warm Up

Name: _____

Date: _____

Convert each temperature in degrees Fahrenheit (F) to degrees Celsius (C). (Formula: °C = °F − 32 ÷ 1.8)

A. 35°F

_____

B. 84°F

_____

C. 90°F

_____

D. 56°F

_____

E. 47°F

_____

Word Problem

F. Zina's dad said she could go swimming when the outside temperature is above 37.8°C. The temperature is 95°F. Can Zina go swimming?

_____

Create your own math problem and explain your solution.

Math Problem:

Explanation/Solution:

_____

Answers: A. 1.67°C B. 28.9°C C. 32.2°C D. 13.3°C E. 8.33°C F. 95°F − 32 ÷ 1.8 = 35°C; No, Zina cannot go swimming.

Warm Up

Name: _____

Date: _____

A **proportion** shows the equality between two numbers. To find the answer, simplify, cross multiply, and divide.

A. $\dfrac{40\%}{100\%} = \dfrac{\$8.00}{n}$

_____

B. $\dfrac{10\%}{100\%} = \dfrac{n}{50}$

_____

C. $\dfrac{3}{4} = \dfrac{n}{10}$

_____

D. $\dfrac{25\%}{100\%} = \dfrac{30}{n}$

_____

Word Problem

Was: $40.00
Now: $28.00

E. Write the math problem to find the sales discount of the tennis shoes.

_____

Create your own math problem and explain your solution.

Math Problem:

Explanation/Solution:

_____

Warm Up

Name: _____

Date: _____

A **proportion** shows the equality between two numbers. To find the answer, simplify, cross multiply, and divide.

A. $\dfrac{10\%}{100\%} = \dfrac{50}{n}$

_____

B. $\dfrac{40\%}{100\%} = \dfrac{50}{n}$

_____

C. $\dfrac{50\%}{100\%} = \dfrac{n}{50}$

_____

D. $\dfrac{20\%}{100\%} = \dfrac{n}{50}$

_____

Word Problem

E. Boot World has a special sale on boots. Which boots have the better price?

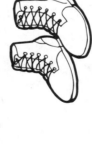
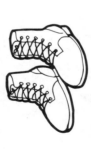

Snake Boots
$40.00 – 25%

Work Boots
Was $40.00 Now $36.00

Create your own math problem and explain your solution.

Math Problem:

_____

Explanation/Solution:

_____

Name: _____ **Date:** _____

Warm Up

The **mean** is the average for a set of numbers. Find the mean for each set of numbers.

A. 65, 96, 72, 28, 93

_____

B. 38, 18, 21, 19, 43

_____

C. 310, 108, 106, 151

_____

D. 127, 145, 194, 367

_____

Word Problem

E. Find the mean number of students.

Grade	K	1st	2nd	3rd	4th	5th	6th
Students	143	104	112	125	96	168	97

Create your own math problem and explain your solution.

Math Problem:

Explanation/Solution: _____

Answers: A. 70.8 B. 27.8 C. 168.75 D. 208.25 E. 121 students

Warm Up

Name: _____

Date: _____

The **mean** is the average for a set of numbers. Find the mean for each set of numbers.

A.

7,865 9,510 3,710 5,492

_____

B.

3,628 6,845 3,343 1,018

_____

C.

17,594 22,598 93,812

_____

D.

72,952 68,710 11,047

_____

Word Problem

E. Find the mean sales price for the homes sold on Park Boulevard.

This Year	$41,928	$64,374	$19,618	$36,737	$58,971
Last Year	$13,810	$63,761	$10,105	$22,410	$79,595

1. What was the mean sales price for last year? _____

2. What is the mean sales price for this year? _____

3. What has happened to the sales prices for the homes over the last year? _____

Create your own math problem and explain your solution.

Math Problem:

Explanation/Solution:

Answers: A. 6,644.25 B. 3,708.5 C. 44,668 D. 50,903 E1. $37,936.20 E2. $44,325.60 E3. The prices have increased.

Name: _____

Date: _____

Warm Up

The **median** is the middle number in a set of numbers in order from smallest to greatest. Write the numbers in order from smallest to greatest. Find the median.

A. 136 175 169

_____ , _____ , _____

The median is _____ .

B. 174 149 184

_____ , _____ , _____

The median is _____ .

C. 110 199 161

_____ , _____ , _____

The median is _____ .

D. 167 143 138

_____ , _____ , _____

The median is _____ .

E. 171 132 183

_____ , _____ , _____

The median is _____ .

Word Problem

F. At the Great Beach Buggy Race, each buggy had a different top speed. Find the mean (average) speed and the median speed for all of the buggies entered in the race. (Remember, mph = miles per

Great Beach Buggy Race			
Dune Buggy	153 mph	Buggy Bee	191 mph
Moon Buggy	194 mph	Sand Buggy	128 mph
Lady Buggy	151 mph		

1. The mean speed is _____ mph.

2. The median speed is _____ mph.

Create your own math problem and explain your solution.

Math Problem: _____

Explanation/Solution: _____

Answers: A. 136, 169, 175; 169 B. 149, 174, 184; 174 C. 110, 161, 199; 161 D. 138, 143, 167; 143 E. 132, 171, 183; 171 F1. 163.4 mph F2. 153 mph

Name: _____

Date: _____

Warm Up

The **median** is the middle number in a set of numbers in order from smallest to greatest. To find the median when there is an even set of numbers, find the two middle numbers, add them together, and divide by two.

A. 15, 58, 32, 92

_____, _____, _____, _____

The median is _____.

B. 73, 21, 75, 15

_____, _____, _____, _____

The median is _____.

C. 36, 25, 91, 18

_____, _____, _____, _____

The median is _____.

D. 78, 34, 96, 22

_____, _____, _____, _____

The median is _____.

Word Problem

E. Round each answer to the nearest penny.

Daily Picnic Specials	
Crunch Chips	$0.77
Super Sodas	$1.20
Cold Sandwich	$1.88
Big Daddy's Cookies	$1.96

1. Find the mean (average) cost of the food items listed on the Daily Picnic Specials board. _____

2. Write the prices in order from least expensive to most expensive. _____, _____, _____, _____

3. Find the median price of the food items. _____

Create your own math problem and explain your solution.

Math Problem:

Explanation/Solution:

_____

Answers: A. 15, 32, 58, 92; 45 B. 15, 21, 73, 75; 47 C. 18, 25, 36, 91; 30.5 D. 22, 34, 78, 96; 56 E1. $1.45 E2. $0.77; $1.20; $1.88; $1.96 E3. $1.54

Name: _____ **Date:** _____

Warm Up

The **mode** is the number that occurs more often than any other number in the set. Find the mode.

A. 93, 98, 95, 93

The mode is _____ .

B. 85, 88, 88, 88

The mode is _____ .

C. 74, 71, 79, 74

The mode is _____ .

D. 62, 65, 60, 60

The mode is _____ .

E. 69, 10, 25, 18, 25, 18, 10, 25

The mode is _____ .

F. 88, 94, 37, 65, 62, 37, 14, 37

The mode is _____ .

G. 195, 133, 142, 186, 141, 142, 133, 133

The mode is _____ .

Word Problem

H. Lance challenges himself to ride his skateboard as many miles as he can in one day. Find the mode and the mean for the number of miles Lance rode his skateboard.

Miles Skateboarded		
14 miles	17 miles	19 miles
14 miles	19 miles	13 miles
18 miles	19 miles	

1. The mode is _____ miles.

2. The mean is _____ miles.

Create your own math problem and explain your solution.

Math Problem: _____

Explanation/Solution: _____

Answers: A. 93 B. 88 C. 74 D. 60 E. 25 F. 37 G. 133 H1. 19 miles H2. 16.625 miles

Name: _____

Date: _____

Warm Up

The **range** is the difference between the lowest number and the highest number in a set of numbers. Write the numbers in order, smallest to greatest. Find the range.

A. 94, 92, 84, 61, 53

____, ____, ____, ____, ____

The range is _____.

B. 84, 41, 23, 81, 46

____, ____, ____, ____, ____

The range is _____.

C. 15, 42, 37, 35, 95

____, ____, ____, ____, ____

The range is _____.

D. 21, 76, 39, 96, 68

____, ____, ____, ____, ____

The range is _____.

Word Problem

E. Find the mean, mode, and range for Samuel's test scores.

Samuel's Test Scores
87%, 87%, 84%, 81%,
83%, 85%, 84%, 87%

1. The mean is _____.

2. The mode is _____.

3. The range is _____.

Create your own math problem and explain your solution.

Math Problem:

_____

Explanation/Solution:

_____

196

Name: _____ **Date:** _____

Warm Up

Identify the number of pairs of congruent sides and parallel sides.

A.

congruent pairs: _____
parallel pairs: _____

B.

congruent pairs: _____
parallel pairs: _____

C.

congruent pairs: _____
parallel pairs: _____

D.

congruent pairs: _____
parallel pairs: _____

Word Problem

E. How are a square and a rectangle similar?

F. How can a square and a rectangle be described so that someone else will know which shape is being described?

Create your own math problem and explain your solution.

Math Problem:

Explanation/Solution: _____

Answers: A. 2, 2 B. 2, 2 C. 2, 2 D. 0, 1 E. They both have 90° angles. They both have 2 pairs of parallel sides, and opposite sides are congruent. F. A square has 4 congruent sides. A rectangle's opposite sides are congruent and parallel.

Name: _____

Date: _____

Warm Up

Find the area for each shape.

Formulas

Triangle
Area = 1/2 (Base x Height)

Square and Rectangle
Area = Length x Width

A.

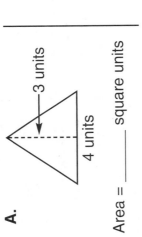

3 units

4 units

Area = _____ square units

B.

3 units

3 units

Area = _____ square units

C.

2 units

5 units

Area = _____ square units

Word Problem

D. Donnie, Ronnie, and Bonnie each drew a shape. Each one was convinced that his or her shape had the greatest area. Determine which one drew the shape with the greatest area.

Donnie's shape

4 units

5 units

_____ square units

Ronnie's shape

4 units

4 units

_____ square units

Bonnie's shape

1 unit

6 units

_____ square units

Create your own math problem and explain your solution.

Math Problem: _____

Explanation/Solution: _____

Answers: A. 6 square units B. 9 square units C. 10 square units D. Donnie's = 10 square units, Ronnie's = 16 square units, Bonnie's = 6 square units; Ronnie's shape has the greatest area.

Warm Up

Name: _____ **Date:** _____

Find the area for each shape.

Formulas

Parallelogram
Area = Base × Height

Trapezoid
Area = 1/2 × Height ×
(Base 1 + Base 2)

A.

3 units

6 units

Area = _____ square units

B.

3 units

2 units

4 units

Area = _____ square units

C.

3 units

4 units

4 units

Area = _____ square units

Word Problem

D. Draw a parallelogram with an area of 24 square units.

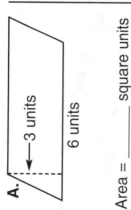

E. Draw a trapezoid with an area of 24 square units.

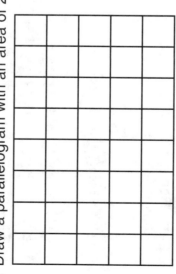

Create your own math problem and explain your solution.

Math Problem:

Explanation/Solution: _____

Answers: A. 18 square units B. 7 square units C. 14 square units D. Sample answer: base of 6, height 4 E. Sample answer: base of 10 and 6, height of 3

199

Warm Up

Name: _____ **Date:** _____

Find the circumference and the area for each circle. Remember, π equals 3.14.

Formula	A.
Circumference	1 unit
$C = \pi \times (2 \times radius)$	$C =$ _____ linear units
Area	$A =$ _____ square units
$A = \pi r^2$	

B.

4 units

$C =$ _____ linear units
$A =$ _____ square units

C.

6 units

$C =$ _____ linear units
$A =$ _____ square units

Word Problem

D. Deandra made a circle with a radius of 3. Find the circumference and the area.

Create your own math problem and explain your solution.

Math Problem: _____

Explanation/Solution:

Answers: A. C = 6.28 linear units; A = 3.14 square units   B. C = 25.12 linear units; A = 50.24 square units   C. 37.68 linear units; A = 113.04 square units
D. C = 18.84 linear units; A = 28.26 square units

Name: _____   **Date:** _____

Warm Up

Find the volume for each solid figure.

Volume
Prism
$V = $ Bases (Length x Width) x Height
Cube
$V = $ Sides^3

A.

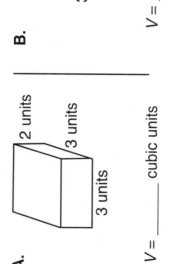

3 units
3 units
2 units

$V = $ _____ cubic units

B.

3 units
3 units
3 units

$V = $ _____ cubic units

C.

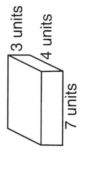

2 units
2 units
2 units

$V = $ _____ cubic units

Word Problem

D. Gilbert needs to find a fish tank for his fish, Freddy. The Fish Shop said Freddy needs to have a tank of at least 80 cubic units. Which fish tank should Gilbert buy?

1.

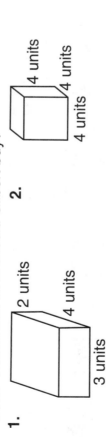

2 units
4 units
3 units

$V = $ _____ cubic units

2.

4 units
4 units
4 units

$V = $ _____ cubic units

3.

3 units
4 units
7 units

$V = $ _____ cubic units

Create your own math problem and explain your solution.

Math Problem:

Explanation/Solution: _____

Answers: A. 18 cubic units B. 27 cubic units C. 8 cubic units D. Gilbert should buy tank #3. D1. 24 cubic units D2. 64 cubic units D3. 84 cubic units

Name: _____ **Date:** _____

Warm Up

Find the volume for each solid figure.

Formulas
Triangle Prism $V = 1/2$ Bases x Height
Pyramid $V = 1/3$ Bases x Height

A.

4 units

4 units   6 units

$V =$ _____ cubic units

B.

4 units

4 units

3 units

$V =$ _____ cubic units

C.

3 units

6 units

5 units

$V =$ _____ cubic units

Word Problem

D. The Pharaoh ordered that a pyramid be made with a volume of at least 60 cubic units and no more than 80 cubic units. What size of pyramid could be made to fit these requirements?

Create your own math problem and explain your solution.

Math Problem:

Explanation/Solution: _____

Answers: A. 48 cubic units B. 16 cubic units C. 30 cubic units D. Sample answer: pyramid with a height of 6 units and bases with a measurement of 6 units each

Name: _____ **Date:** _____

Warm Up

Find the volume for each cone.

Cone
1. Find the area of the circle. $A = \pi r^2$
2. Multiply the area by the height. $V = \dfrac{bh}{3}$
3. Divide by 1/3.

A.

12 units

$r = 3$ units

$V =$ _____ cubic units

B.

12 units

$r = 5$ units

$V =$ _____ cubic units

Word Problem

C. There are two sizes of soccer cones. One size has a radius of 2 units and a height of 12 units. The other size has a radius of 4 units and a height of 8 units. Which one has a volume of about 50 cubic units?

Create your own math problem and explain your solution.

Math Problem: _____

Explanation/Solution: _____

Answers: A. 113.04 cubic units B. 314 cubic units C. 1st cone 50.24 cubic units, 2nd cone 133.97 cubic units—The 1st cone has volume of about 50 cubic units.

Name: _____

Date: _____

Warm Up

Find the volume for each cylinder.

Cylinder
1. Find the area of the circle. $A = \pi r^2$
2. Multiply the area by the height of the cylinder.

A.

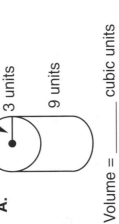

3 units

9 units

Volume = _____ cubic units

B.

2 units

10 units

Volume = _____ cubic units

Word Problem

C. A soda can has a height of 8 units and a radius of 4 units. A frozen orange juice can has a height of 6 units and a radius of 3 units. Which can has the greater volume?

Create your own math problem and explain your solution.

Math Problem:

Explanation/Solution:

Answers: A. 254.34 cubic units B. 125.6 cubic units C. soda can = 401.92 cubic units; orange juice can = 169.56; The soda can has the greater volume.

Warm Up

Name: _____

Date: _____

Identify the end points.

A.

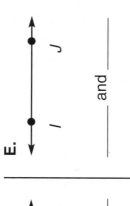

A   B

_____ and _____

B.

C   D

_____ and _____

C.

E   F

_____ and _____

D.

G   H

_____ and _____

E.

I   J

_____ and _____

Word Problem

F.

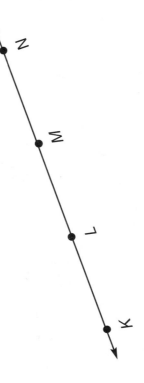

1. Identify the end points.

   _____ and _____

2. Identify each line segment two ways.

   _____ or _____

   _____ or _____

   _____ or _____

Create your own math problem and explain your solution.

Math Problem:

Explanation/Solution:

_____

Name: _____ **Date:** _____

Warm Up

Identify each line two ways.

A.

A   B

_____ or _____

B.

C   D

_____ or _____

C.

E   F

_____ or _____

D.

G   H

_____ or _____

E.

I   J

_____ or _____

Word Problem

F. Identify each line segment two ways.

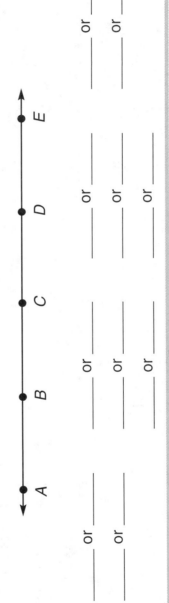

A   B   C   D   E

_____ or _____

_____ or _____

_____ or _____

_____ or _____

_____ or _____

_____ or _____

_____ or _____

_____ or _____

_____ or _____

_____ or _____

Create your own math problem and explain your solution.

Math Problem:

Explanation/Solution: _____

Answers: A. $\overleftrightarrow{AB}$ or $\overleftrightarrow{BA}$ B. $\overleftrightarrow{CD}$ or $\overleftrightarrow{DC}$ C. $\overleftrightarrow{EF}$ or $\overleftrightarrow{FE}$ D. $\overleftrightarrow{GH}$ or $\overleftrightarrow{HG}$ E. $\overleftrightarrow{IJ}$ or $\overleftrightarrow{JI}$ F. $\overline{AB}$ or $\overline{BA}$, $\overline{AC}$ or $\overline{CA}$, $\overline{AD}$ or $\overline{DA}$, $\overline{AE}$ or $\overline{EA}$, $\overline{BC}$ or $\overline{CB}$, $\overline{BD}$ or $\overline{DB}$, $\overline{BE}$ or $\overline{EB}$, $\overline{CD}$ or $\overline{DC}$, $\overline{CE}$ or $\overline{EC}$, $\overline{DE}$ or $\overline{ED}$

Name: _____   Date: _____

Warm Up

Identify the end points.

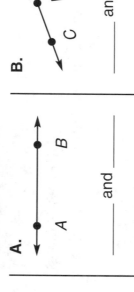

A. _____ and _____

B. _____ and _____

C. _____ and _____

D. _____ and _____

E. _____ and _____

Word Problem

F. What are "end points"? _____

G. What is a line segment? _____

H. In the empty space, draw a line segment with the end points *R* and *S*.

1. On $\overline{RS}$, add the line segments *TUV*.

2. What are the end points? _____

3. What are the line segments? _____

Create your own math problem and explain your solution.

Math Problem:

Explanation/Solution:

Answers: A. A and B B. C and E C. F and I D. J and M E. N and Q F. The beginning and ending points on a line. G. A line segment is a section of the line.
H2. R and S H3. $\overline{RS}$ or $\overline{SR}$, $\overline{RT}$ or $\overline{TR}$, $\overline{RU}$ or $\overline{UR}$, $\overline{RV}$ or $\overline{VR}$, $\overline{ST}$ or $\overline{TS}$, $\overline{SU}$ or $\overline{US}$, $\overline{SV}$ or $\overline{VS}$, $\overline{TU}$ or $\overline{UT}$, $\overline{TV}$ or $\overline{VT}$, $\overline{UV}$ or $\overline{VU}$

Warm Up

Name: _____ Date: _____

Identify each problem as a *line*, a *line segment*, or a *ray*.

A.

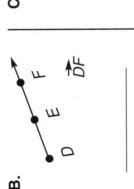

B ● —— ● C

$\overline{BC}$

_____

B.

D ●
  E ●
    F ●↗

$\overrightarrow{DF}$

_____

C.

↖● G
   ● H
    ● I↗

$\overleftrightarrow{GI}$

_____

D.

J ●
  K ●

$\overline{JK}$

_____

E.

L ●
  M ●↑

$\overrightarrow{LM}$

_____

Word Problem

F. How is a ray like a line? _____

_____

G. How is a ray different from a line? _____

_____

Create your own math problem and explain your solution.

Math roblem: _____

Explanation/Solution: _____

Answers: A. line segment B. ray C. line D. line segment E. ray F. They both go on endlessly. G. A ray goes on endlessly in one direction. A line goes on endlessly in two directions.

Name: _____ **Date:** _____

Warm Up

Identify each set of lines as *parallel* (||), *perpendicular* ($\perp$) or *intersecting*.

A.

A B
C D

B.

G
E ——— F
H

C.

I J
K L

D.

M P
O N

E.

R
P ——— Q
S

Word Problem

F.

1. Draw $\overleftrightarrow{AB}$ || to $\overleftrightarrow{CD}$.

2. Draw $\overleftrightarrow{EF}$ $\perp$ to $\overleftrightarrow{GH}$.

3. Draw $\overleftrightarrow{IJ}$ intersecting $\overleftrightarrow{KL}$.

Create your own math problem and explain your solution.

Math Problem:

_____

Explanation/Solution:

Warm Up

Name: _____ **Date:** _____

Identify each angle two ways.

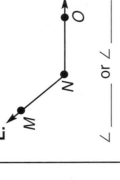

A.

∠ _____ or ∠ _____

B.

∠ _____ or ∠ _____

C.

∠ _____ or ∠ _____

D.

∠ _____ or ∠ _____

E.

∠ _____ or ∠ _____

Word Problem

F. Draw ∠PQR = 90°

∠PQR = 90°

G. Draw ∠STU with a narrower angle than ∠PQR.

∠STU = _____ °

H. Draw ∠VWX with a wider angle than ∠PQR.

∠VWX = _____ °

Create your own math problem and explain your solution.

Math Problem:

Explanation/Solution:

Answers: A. ∠ABC or ∠CBA B. ∠DEF or ∠FED C. ∠GHI or ∠IHG D. ∠JKL or ∠LKJ E. ∠MNO or ∠ONM F. ∠PQR = 90° G. ∠STU is narrower than 90°. H. ∠VWX has an angle larger than 90°.

Name: _____   **Date:** _____

Warm Up

Use a protractor to measure each angle.

A.

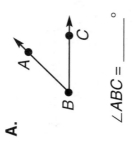

∠ABC = _____ °

B.

∠DEF = _____ °

C.

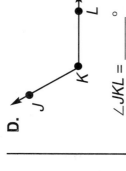

∠GHI = _____ °

D.

∠JKL = _____ °

E.

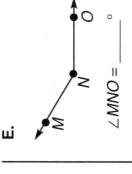

∠MNO = _____ °

Word Problem

F. Read the clues to find the angle each student made.

- Jason did not draw the widest angle nor the narrowest angle.
- Marilyn's angle is wider than Jason's angle.

∠PQR = 45°

∠STU = 90°

∠VWX = 150°

1. Marilyn's angle is _____ .

2. Jason's angle is _____ .

3. Sammy's angle is _____ .

Create your own math problem and explain your solution.

Math Problem:

Explanation/Solution:

Warm Up

Name: _____ **Date:** _____

Measure and identify each angle. **Right angles** are 90°. **Acute angles** are less than 90°. **Obtuse angles** are more than 90°.

A.

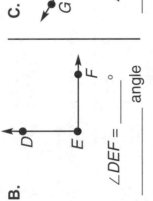

∠ABC = _____° _____ angle

B.

∠DEF = _____° _____ angle

C.

∠GHI = _____° _____ angle

D.

∠JKL = _____° _____ angle

E.

∠MNO = _____° _____ angle

Word Problem

F. Measure and identify each angle.

Angle	∠ABC	∠BCE	∠CED	∠EDA	∠DAB
Degrees					
Type					

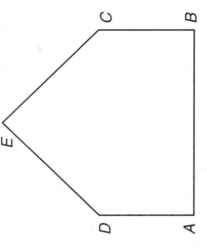

Create your own math problem and explain your solution.

Math Problem: _____

Explanation/Solution:

_____

Answers: A. 30°, acute B. 90°, right C. 150°, obtuse D. 10°, acute E. 95°, obtuse F. ∠ABC = 90°, right; ∠BCE = 135°, obtuse; ∠CED = 90°, right; ∠EDA = 135°, obtuse; ∠DAB = 90°, right

213

Warm Up

To bisect an angle means to divide it in half. An **angle bisector** is a ray that separates an angle into two congruent angles. Identify the angle bisector.

A.

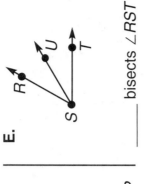

_____ bisects ∠ABC

B.

_____ bisects ∠FGH

C.

_____ bisects ∠JKL

D.

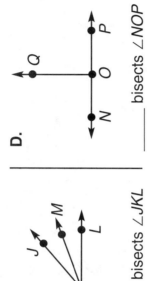

_____ bisects ∠NOP

E.

_____ bisects ∠RST

Word Problem

F. Marlene made ∠FGH with an opening of 42°. She bisected ∠FGH with $\overrightarrow{GI}$.

1. Identify the angle bisector. _____

2. Identify the adjacent angles. _____

3. Identify the type of angle for ∠FGH. _____

Create your own math problem and explain your solution.

Math Problem:

Explanation/Solution: _____

Name: _____

Date: _____

Warm Up

Identify each pair of angles as *adjacent* or *not adjacent*.

A.

∠ABC and ∠CBD   adjacent   not adjacent

B.

∠EFH and ∠EFG   adjacent   not adjacent

C.

∠IJK and ∠LJK   adjacent   not adjacent

D.

∠MNP and ∠PNO   adjacent   not adjacent

Word Problem

E. Complimentary angles are two angles with a sum of 90°. **Supplementary angles** are two angles with a sum of 180°. Measure and identify each pair of angles.

1.

∠ABD and ∠DBC

2.

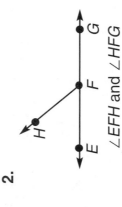

∠EFH and ∠HFG

Create your own math problem and explain your solution.

Math Problem: _____

Explanation/Solution: _____

Warm Up

Identify the adjacent angles.

A.

_____ and _____ are adjacent.

B.

_____ and _____ are adjacent.

C.

_____ and _____ are adjacent.

D.

_____ and _____ are adjacent.

E.

_____ and _____ are adjacent.

Word Problem

F. Look at the angles.

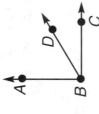

1. Write the adjacent angles. _____
2. Write the non-adjacent angles. _____

Create your own math problem and explain your solution.

Math Problem: _____

Explanation/Solution: _____

Answers: A. ∠BAD and ∠DAC are adjacent angles. B. ∠EFH and ∠HFG are adjacent angles. C. ∠IJL and ∠LJK are adjacent angles. D. ∠MNP and ∠PNO are adjacent angles. E. ∠QRT and ∠TRS are adjacent angles. F1. ∠ABD and ∠DBC are adjacent. F2. ∠ABC is not adjacent to ∠DBA nor to ∠DBC.

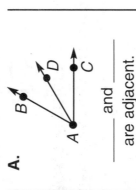

Name: _____

Date: _____

Warm Up

Measure each angle. Identify each angle as "congruent" (=) or "not congruent" (≠).

A.

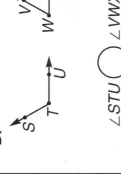

∠ABC ◯ ∠DEF

B.

∠GHI ◯ ∠JKL

C.

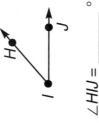

∠MNO ◯ ∠PQR

D.

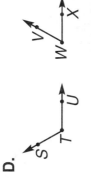

∠STU ◯ ∠VWX

Word Problem

E. Congruent angles are two angles with the same measurement. Find the measurement of each angle. Then find the angles that are congruent.

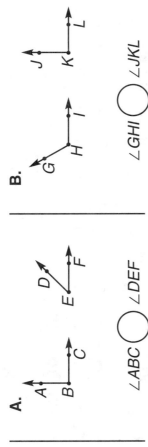

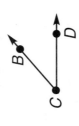

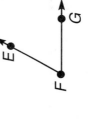

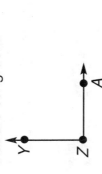

∠YZA = _____ °   ∠BCD = _____ °   ∠EFG = _____ °   ∠HIJ = _____ °

Create your own math problem and explain your solution.

Math Problem: _____

Explanation/Solution: _____

Answers: A. ∠ABC ≠ ∠DEF B. ∠GHI ≠ ∠JKL C. ∠MNO = ∠PQR D. ∠STU ≠ ∠VWX E. ∠YZA = 90°, ∠BCD = 40°, ∠EFG = 60°, ∠HIJ = 40°; The congruent angles are ∠BCD and ∠HIJ.

Name: _____

Date: _____

Warm Up

All of the inside angles of a triangle equal 180°. Identify the measurement of the missing angle.

A.

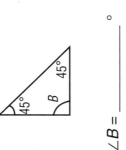

∠A = _____ °

B.

∠B = _____ °

C.

∠C = _____ °

Word Problem

D. A triangle can be classified by its angles and by its number of congruent sides. Classify Toby's three triangles two different ways.

Congruent Sides	Angles
0 = scalene	3 angles less than 90° = acute
2 = isosceles	one 90° angle = right
3 = equilateral	angle more than 90° = obtuse

1.

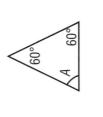

2.

3.

Sides: _____ _____ _____

Angles: _____ _____ _____

Create your own math problem and explain your solution.

Math Problem: _____

Explanation/Solution: _____

218

Name: _____

Date: _____

Warm Up

All of the inside angles of a 4-sided shape equal 360°. Find the missing angle for each shape.

A.

90° A
90°   90°

∠A = _____°

B.

100° 100°
B   80°

∠B = _____°

C.

120°
30°   30°
C

∠C = _____°

D.

70° D
110°   70°

∠D = _____°

Word Problem

E. Dora made the 4-sided shape. She measured each angle. Are her measurements correct? Why or why not?

40°

40°

40°

40°

_____

_____

_____

Create your own math problem and explain your solution.

Math Problem:

Explanation/Solution: _____

219

Name: _____ **Date:** _____

Warm Up

Draw a line under the base. Circle the exponent.

A. 8^5

B. 4^9

C. 9^6

D. 6^7

E. 4^2

F. 8^9

G. _____

H. _____

I. 4^3

J. 2^6

G. 9^5

H. 4^8

Word Problem

K. How is 2 x 3 different from 2^3?

Create your own math problem and explain your solution.

Math Problem:

Explanation/Solution: _____

Answers: A. base: 8, exponent: 5 B. base: 4, exponent: 9 C. base: 9, exponent: 6 D. base: 6, exponent: 7 E. base: 4, exponent: 2 F. base: 8, exponent: 9 G. base: 9, exponent: 5 H. base: 4, exponent: 8 I. base: 4, exponent: 3 J. base: 2, exponent: 6 K. 2 x 3 = 6 and 2^3 is 2 x 2 x 2 = 8

Name: _____

Date: _____

Warm Up

Rewrite each number in exponential form.

A.

$4 \times 4 \times 4 \times 4 \times 4 \times 4$

_____

B.

$10 \times 10 \times 10 \times 10$

_____

C.

$9 \times 9 \times 9 \times 9 \times 9 \times 9$

_____

D.

$7 \times 7 \times 7 \times 7 \times 7$

_____

Word Problem

E. Serena has 4^3 trading cards. Christopher has 3^4 trading cards. Who has more trading cards?

Create your own math problem and explain your solution.

Math Problem: _____

Explanation/Solution: _____

Answers: A. 4^7 B. 10^4 C. 9^6 D. 7^5 E. Christopher has more trading cards. (Serena has 64 trading cards. Christopher has 81 trading cards.)

221

Name: _____ **Date:** _____

Warm Up

Write the value.

A. 4^5 _____ B. 6^4 _____ C. 6^2 _____

D. 8^3 _____ E. 3^3 _____

Word Problem

F. Complete the table to the right. Then circle the larger number in each pair below.

n	n^2	n^3	n^4	n^5	n^6
2	4				
3	9				
4	16				
5	25				

1. 2^5 4^2

2. 3^3 4^4

3. 3^4 2^5

4. 4^3 5^2

5. 5^3 2^6

6. 2^6 5^2

Create your own math problem and explain your solution.

Math Problem:

_____

Explanation/Solution:

_____

Name: _____

Date: _____

Warm Up

Write each number in standard form.

A. _____
2^4

B. _____
9^3

C. _____
6^2

D. _____
7^3

E. _____
8^3

Word Problem

F. Eileen has 8^2 sheep. Felix has 6^3 sheep. Who has more sheep?

Create your own math problem and explain your solution.

Math Problem:

Explanation/Solution:

_____

Name: _____

Date: _____

Warm Up

Circle the variable.

A. $10 - n$	**C.** $2 + e$	**E.** $3 - g$	**G.** $h + 3$	**I.** $15 - k$
B. $9 + x$	**D.** $f + 8$	**F.** $11 - a$	**H.** $d - 2$	**J.** $31 + m$

Word Problem

K. Write the math problem using a variable.

Casey made 31 model airplanes. Trevor made model airplanes too. Now there are 65 model airplanes. How many airplanes did Trevor make?

Create your own math problem and explain your solution.

Math Problem:

Explanation/Solution: _____

224

Name:

Date:

Warm Up

Write the equation. Solve the problem. Use *n* for the variable.

A. Yoshiko had $89.91. She bought a fish bowl and was given $18.27 in change. How much did the fish bowl cost?

_____

B. Damian and Julian made 82 origami animals. Damian made 19 of them. How many did Julian make?

_____

C. Ed had $22.10. He earned money by doing chores around the house. He now has $31.43. How much money did he earn?

_____

Word Problem

D. Dorian is thinking of a mystery number. If 39 is added to the number, the answer is 83. What is Dorian's mystery number?

Create your own math problem and explain your solution.

Math Problem:

Explanation/Solution:

_____

Answers: A. $89.91 – n = $18.27, $89.71 – $18.27 = n, n = $71.64 B. 19 + n = 82, n = 82 – 19, n = 63 C. $22.10 + n = $31.43, n = $31.43 – $22.10, n = $9.33 D. n + 39 = 83, n = 83 – 39, n = 44

Name: _____ **Date:** _____

Warm Up

Write the equation. Solve the problem. Use *n* for the variable.

A. J.C. had 37 nails and some bolts. He had 83 items in all. How many bolts did J.C. have?

_____

B. Honey had 59 pennies and some nickels. She had 91 coins in all. How many nickels did Honey have?

_____

C. Francine had some pretzels and 57 peanuts. She had 83 items in all. How many pretzels did Francine have?

_____

Word Problem

D. Laurel made 53 birdhouses. She sold some of them and now has 25 birdhouses left. How many birdhouses did Laurel sell?

_____

Create your own math problem and explain your solution.

Math Problem:

Explanation/Solution:

_____

Answers: A. 37 + n = 83, n = 83 − 37, n = 46 bolts B. 59 + n = 91, n = 91 − 59, n = 32 nickels C. n + 57 = 83, n = 83 − 57, n = 26 pretzels D. 53 − n = 25, n = 53 − 25, n = 28

226

Name: _____

Date: _____

Warm Up

Write the algebraic expression using z for the variable.

A. 16 decreased by the number z

_____

B. 67 multiplied by the number z

_____

C. the sum of 43 and the number z

_____

D. ten less than the number z

_____

E. 6 more than the number z

_____

Word Problem

F. Write the algebraic expression using z for the variable (unknown quantity).

Robert made 11 more hot dogs than Paula. How many hot dogs did Robert make?

Create your own math problem and explain your solution.

Math Problem:

Explanation/Solution:

Answers: A. 16 − z B. 67 × z C. 43 + z D. z − 10 E. z + 6 F. z + 11

Name: _____

Date: _____

Warm Up

Write the algebraic expression for each word problem. Use z for the variable.

A. Martha is ten years older than Ben.

_____

B. Kathy made twice as many placemats than Christine.

_____

C. Jinks caught 54 fewer fish than Pam.

_____

D. Ken has 25¢ less than Cieta.

_____

Word Problem

E. Identify each of the equations as "true," "false," or "open." Answer the question.

1. $10 - 7 = 9$	**2.** $9 - 4 = n$	**3.** $10 + 1 = 11$	**4.** $2 + n = 7$	**5.** $5 + 8 = 12$
_____	_____	_____	_____	_____

6. What makes a true equation different from a false equation?

_____

Equations
True: $15 - 15 = 0$
False: $1 + 1 = 9$
Open: $n - 2 = 6$

Create your own math problem and explain your solution.

Math Problem: _____

Explanation/Solution:

227

Name: _____

Date: _____

Warm Up

Circle the math problem.

A. Hal counted 672 beads. Cal counted 791 beads. How many beads were counted in all?

$672 + 791 = n$ $672 - 791 = n$

$672 \times 791 = n$ $672 \div 791 = n$

B. Nell saw 80 snails. Bea saw 9 times more snails than Nell. How many snails did Bea see?

$80 + 9 = n$ $80 - 9 = n$

$80 \times 9 = n$ $80 \div 9 = n$

C. Fred counted 30 bees. Ed counted half as many bees as Fred. How many bees did Ed count?

$30 + 2 = n$ $30 - 2 = n$

$30 \times 2 = n$ $30 \div 2 = n$

Word Problem

D. Hank saw 44 ladybugs eating aphids. Frank saw 21 more ladybugs than Hank. How many ladybugs did Frank see?

Create your own math problem and explain your solution.

Math Problem:

Explanation/Solution:

Answers: A. $672 + 791 = n$ B. $80 \times 9 = n$ C. $30 \div 2 = n$ D. $44 + 21 = 65$ ladybugs

Name: _____

Date: _____

Warm Up

Let $n = 7$. Solve.

A.

$25 + n =$ _____

B.

$31 \times n =$ _____

C.

$98 \div n =$ _____

D.

$37 - n =$ _____

E.

$(37 - n) \times n =$ _____

Word Problem

F. Write the equation. Solve.

Seth scored a total of 53 points at target practice. Carrie scored 91 more points than Seth did. Thomas scored half as many points than Carrie. How many points did Thomas score?

Create your own math problem and explain your solution.

Math Problem:

Explanation/Solution: _____

Answers: A. 32 B. 217 C. 14 D. 30 E. $(30) \times 7 = 210$ F. $(53 + 91) \div 2 = 72$ points

Warm Up

Name: _____

Date: _____

Complete each table.

A.

n	6	10	7	8
n − 4	2			

B.

n	9	8	14	10
n + 5	14			

C.

n	3	5	2	1
n × 3	9			

Word Problem

D. Write the algebraic expression.

Candy jumped her skateboard for a long distance. Randy jumped his skateboard 37 fewer feet than Candy.

1. How far did Candy and Randy jump their skateboards? _____

2. If Candy jumped her skateboard 93 feet, how long was Randy's jump? _____

Create your own math problem and explain your solution.

Math Problem:

Explanation/Solution:

Answers: A. 6, 3, 4 B. 13, 19, 15 C. 15, 6, 3 D1. n − 37 D2. 93 − 37 = 56 feet

Name: _____ **Date:** _____

Warm Up

Replace the variable z with the number 4. Rewrite and solve each problem.

A. z x 9

_____

B. z ÷ 2

_____

C. z + 8

_____

D. z – 2

_____

Word Problem

E. Find the number for the variable that will make all math sentences true.

	+ 6	– 6	x 2	÷ 3
z	15	3	18	3

Create your own math problem and explain your solution.

Math Problem:

Explanation/Solution: _____

Name: _____ **Date:** _____

Warm Up

Do the operation in parentheses first. Let $n = 10$.

A.

$(76 + n) - 56 =$

_____

B.

$(14 - n) + 18 =$

_____

C.

$(11 - n) + 65 =$

_____

D.

$(10 + 38) - n =$

_____

E.

$(52 - 10) + n =$

_____

Word Problem

F. Write the equation for each question. Then solve.

1. How many more animals were in this year's fair than in last year's fair? _____

2. If 362 of the animals in the fair this year were adult animals, how many were still piglets and lambs? _____

County Fair

	Pigs	Sheep
This year:	687	937
Last year:	416	714

Create your own math problem and explain your solution.

Math Problem:

_____

Explanation/Solution:

_____

Answers: A. $86 - 56 = 30$ B. $4 + 18 = 22$ C. $1 + 65 = 66$ D. $48 - 10 = 38$ E. $42 + 10 = 52$ F1. $(687 + 937) - (416 + 714) = 1,624 - 1,130 = 494$ more animals F2. $(687 + 937) - 362 = 1,624 - 362 = 1,262$ were still piglets and lambs.

Name: _____

Date: _____

Warm Up

Find the value of x.

A.
$$x = 5 \times 2$$

B.
$$x = 4^2 + 3$$

C.
$$x = (10 - 3) - 2^2$$

D.
$$x = (81 \div 9) + 9$$

E.
$$x = 4^2 - (10 + 6)$$

F.
$$x = 10 \div 5$$

G.
$$x = 6 \div 3$$

H.
$$x = (9 + 8) - (3 \times 4)$$

I.
$$x = 2 + 3 + 2 + 3$$

J.
$$x = 100^2 \times 0$$

Word Problem

K. When planning a party, it is important to purchase the correct amount of food. Help the party planners decide how many supplies to order. Write the math equation that answers each question. (Round each answer to the nearest whole number.)

1. Purchase 9 ounces of punch for each guest.
2. Buy 1/6 of a pound of meat for each guest.
3. Order twice the number of napkins as guests.
4. Have 1.3 chairs for each guest.
5. Have 1/4 of a pound of candy for each guest.
6. Buy one gift for each family of guests.

Guest List	Numbers
The Browns	8
The Greens	4
The Blacks	10
The Whites	1

Create your own math problem and explain your solution.

Math Problem: _____

Explanation/Solution: _____

Answers: A. x = 10 B. x = 19 C. x = 3 D. x = 18 E. x = 0 F. x = 2 G. x = 2 H. x = 5 I. x = 10 J. x = 0 K1. 23 x 9 = 207 ounces K2. 23 x 1/6 = 3 5/6 pounds or 4 pounds K3. 23 x 2 = 46 K4. 23 x 1.3 = 29.9 or 30 chairs K5. 23 x 1/4 = 5 3/4 or 6 pounds of candy K6. 4 x 1 = 4 gifts

234

Warm Up

Find the value of the variable y.

A.

$(2 + y) - 9 = 3$

_____

B.

$31 + (y - 5) = 33$

_____

C.

$(35 + y) - 4 = 35$

_____

D.

$(96 - 59) + y = 73$

_____

E.

$y - 96 = 14$

_____

Word Problem

F. Brynn had 71 boxes of Yummy Delights and some boxes of Gooey Chewies. She sold 25 boxes of cookies and now has 49 boxes of cookies left. How many boxes of Gooey Chewies did Brynn have?

Create your own math problem and explain your solution.

Math Problem:

Explanation/Solution: _____

Answers: A. $y = 10$ B. $y = 7$ C. $y = 4$ D. $y = 36$ E. $y = 110$ F. $(71 + y) - 25 = 49$, $y = 3$

Name: _____

Date: _____

Warm Up

A. _____

B. _____

C. _____

D. _____

E. _____

Word Problem

F.

Create your own math problem and explain your solution.

Math Problem:

Explanation/Solution:

Answers: _____

Time

second

minute = 60 seconds

hour = 60 minutes

quarter past

quarter till

half past

day = 24 hours

week = 7 days

month = 28 or 30 or 31 days

year = 365 days

Perimeter

The *perimeter* is the area around a given shape. Add the length of all sides to find the perimeter.

Example:

Math Vocabulary

Name: _____

Addition (+)

Addition is when two or more addends (numbers) are grouped together to equal a sum (answer).

Example:

Money

Write each coin's name and value on the line.

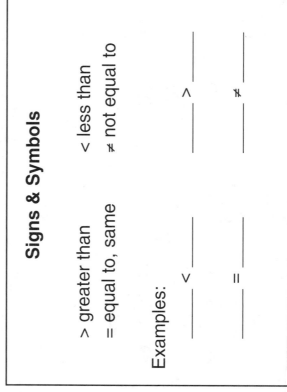

_____

_____

_____   _____   _____

Signs & Symbols

> greater than   < less than

= equal to, same   ≠ not equal to

Examples:

<   _____   >   _____

=   _____   ≠   _____

Subtraction (−)

Subtraction is when a given number of items is removed from the group. The remaining number of items is the answer (difference).

Example:

Money

Write the value of each paper bill on the line.

_____   _____

_____   _____

Ratio

A *ratio* tells the mathematical relationship between two items. The ratio can be written two ways—using (:) or (/) to separate the two numbers.

Example:

Exact & Estimate

Exact means that the answer is a precise number.

Example of an exact answer: _____

Estimate means that the answer is an approximate number.

Example of an estimated answer: _____

Money

Money can be recorded in different ways.

Examples:

Standard form: _____

Cents sign: _____

Fraction of $1: _____

Percent of $1: _____

Words: _____

Order of Operations

The *order of operation* tells which step to do first when solving a math problem using multiple symbols.

1. Parenthesis
2. Exponents
3. Multiplication
4. Division
5. Addition
6. Subtraction

Probability

Probability means the likelihood of an event happening or not happening.

Example of something that has a strong chance of happening: _____

Example of something that has a poor chance of happening: _____

Fractions

A *fraction* is a part of a whole.

• The top number is the numerator. The numerator tells how many pieces or segments of the whole are being used.

• The bottom number is the denominator. The denominator tells the total number of pieces or segments in the item.

Example:

Rounding Numbers

When rounding numbers, look at the number to the right of the place the number is being rounded to. If the number is 5 or larger, round up to the next whole number. If the answer is 4 or less, round down to the previous whole number.

Example of rounding up: _____
Example of round down: _____

Addition & Subtraction Fact Families

A *fact family* is a set of numbers that are used to create two addition and two subtraction problems.

Example:

Place Value

Place value refers to the value given to a digit based on its position within a specific number.

Example:

Skip Counting

Skip counting uses specific numbers within a given range, usually of the same value.

Examples: Counting dimes by tens and nickels by fives.

Other examples of skip counting:

Adding & Subtracting Fractions with Like Denominators

Add or subtract the numerator as you would when solving an addition or subtraction problem using whole numbers. The denominator does not change.

Example:

Ordering Numbers

To order numbers from smallest to greatest, look at the digit in the largest place. The number with the lowest digit is first. The number with the highest digit is last.

Example:

Factors

Factors are the numbers that can be used to multiply to equal a given product.

Example:

Metric System & Conversion Chart

The *metric system* is a way of measuring items around the idea of "ten."

millimeter (mm)
centimeter (cm) = 10 mm
meter (m) = 100 cm or 1,000 mm
cm = .39 inches
m = 1.094 yards
$°F = °C \times 1.8 + 32$

Multiplication

Multiplication is just repeated addition—multiplying a given number by the number or times it is used within a problem.

Example:

Division

Division is just repeated subtraction—subtracting a given quantity from a specific number over and over again until all of the items have been used or there are too few left to make a whole set. The ones left are the remainder.

Example:

Equivalent Fractions

Equivalent fractions are fractions that have the same value.

Example:

Multiplication & Division Fact Families

Multiplication and division fact families use a set of numbers to make two multiplication and two division problems.

Example:

United States Standard Measurement & Conversion Chart

inches (in.)

foot (ft.) = 12 inches

yard (yd.) = 36 inches or 3 feet

mile = 1.609 km

°C = °F − 32 ÷ 1.8

teaspoon (tsp.)

tablespoon (T) = 3 tsp.

quart (qt.)

gallon (g) = 4 quarts

pound (lb.) = 16 ounces (oz.)

ton = 2,000 lb.

Simplifying a Fraction

The numerator and the denominator in a fraction are divided by the same number.

Example:

Improper Fraction

An *improper fraction* has a numerator that is larger than the denominator.

Example:

Multiplying Fractions

Multiply the numerators and multiply the denominators. Write the answer in its simplest form.

Example:

Unlike Denominators

When working with two or more fractions with different denominators, find the common denominator for all fractions and multiply each fraction by the specific number to reach the new denominator. Then add or subtract.

Example:

Mixed Fraction

A *mixed fraction* has both a whole number and a fraction as part of its set.

Example:

Percent

A *percent* is a part, or fraction, of a whole amount. The percent (%) symbol is to represent the fractional amount.

Example:

Composite Number

A *composite number* is a number that can be divided by more than two numbers.

Example:

Decimal

A *decimal* is a part, or fraction, of a whole amount. A decimal point is used to separate the ones column from the tenths column.

Example:

Prime Number

A *prime number* is a number that can only be divided by the number one and itself. (Two is considered to be a prime number.)

Example:

Mean

The *mean* is the average for a set of numbers.

Example:

Median

The *median* is the middle number in a set of numbers ordered from smallest to greatest.

Example:

Average

An *average* tells the number of items that can be given in equal numbers. To find the average, add all of the numbers together then divide by the numbers in the set.

Example:

Proportion

A *proportion* shows the equality between two numbers. To find the answer, simplify, cross multiply, and divide.

Example:

Range

The *range* is the difference between the smallest and largest number within a set of numbers.

Example:

Formula for Finding the Area of a Square and Rectangle

Area = length x width

Example:

Mode

The *mode* is the number that occurs more often than any other number in the set.

Example:

Formula for Finding the Area of a Triangle

Area = 1/2 (Base x height)

Example:

**Formula for Finding
the Area of a Trapezoid**

Area = 1/2 x height x (base 1 + base 2)

Example:

**Formula for Finding
the Area of a Circle**

pi or π = 3.14

Area = pi x r^2

$A = \pi r^2$

Example:

**Formula for Finding
the Area of a Parallelogram**

Area = Base x height

Example:

**Formula for Finding
the Circumference of a Circle**

pi or π = 3.14

Circumference = pi x (2 x radius)

$C = \pi \times (2r)$

Example:

**Formula for Finding
the Volume of a Cube**

Volume = side x side x side

$V = s^3$

Example:

**Formula for Finding
the Volume of a Pyramid**

Volume = 1/3 Bases x height

Example:

**Formula for Finding
the Volume of a Rectangular Prism**

Volume = Bases (length x width) x height

$V = Bh$

Example:

**Formula for Finding
the Volume of a Triangular Prism**

Volume = 1/2 Bases x height

Example:

Formula for Finding the Volume of a Cylinder

1. Find the area of the circle. $\quad A = \pi r^2$
2. Multiply the area by the height of the cylinder. $\quad V = A \times h$

Formula for Finding the Volume of a Cone

1. Find the area of the circle. $\quad A = \pi r^2$
2. Multiply the area by the height. $\quad V = \dfrac{Bh}{3}$
3. Divide by 1/3.

Example:

Line Segment

A *line segment* is part of a line. The line segment can be named two different ways.

Example:

Lines

End points are found at the ends of a line.

A line can go on indefinitely in opposite directions.

Example:

Parallel Lines

Parallel lines are two lines that run side by side maintaining an equal separation from beginning to end. Parallel lines are described with the symbol ||.

Example:

Intersecting Lines

Intersecting lines cross each other at some point.

Example:

Rays

A *ray* is a line that extends indefinitely in one direction.

Example:

Perpendicular Lines

Perpendicular lines are two lines that intersect to form a 90° angle. Perpendicular lines are described with the symbol ⊥.

Example:

Angle Bisector

An *angle bisector* is a ray that divides an angle exactly in half. The bisecting ray is identified by the vertex and its ending point.

Example:

Complimentary Angle

Complimentary angles are two angles with a sum of 90°.

Example:

Angles

Angles are described by using the two endpoints and the vertex. (∠ means angle).

• obtuse angle—any angle larger than 90°
• acute angle—any angle smaller than 90°
• right angle—a 90° angle

Example of the different kinds of angles:

Adjacent Angles

Adjacent angles share a common vertex and a common ray.

Example:

Congruent Angles

Congruent angles are two angles that are the same in size.

= congruent   ≠ not congruent

Examples:

Triangles

A triangle can be classified by its angles.

Acute = contains three angles less than 90°
Right = contains one 90° angle
Obtuse = contains an angle greater than 90°

Examples:

Supplementary Angles

Supplementary angles are two angles with a sum of 180°.

Example:

Triangles

Triangles can be classified by the number of congruent sides.

Scalene = 0 congruent sides
Isosceles = 2 congruent sides
Equilateral = 3 congruent sides

Examples:

Variable

A *variable* is used to identify an unknown number or quantity within a specific equation

Example:

Example:

Exponent

Exponents are used to tell how many times to multiply a number (base) by itself (exponent).

Example:

Example: